STARGAZING

www.royalcollins.com

STARGAZING

EXPLORING THE MYSTERIES OF
THE STARLINK

KEVIN CHEN

RC

Books Beyond Boundaries

ROYAL COLLINS

Stargazing: Exploring the Mysteries of the Starlink

Kevin Chen

First published in 2024 by Royal Collins Publishing Group Inc.
Groupe Publication Royal Collins Inc.
550-555 boul. René-Lévesque O Montréal (Québec)
H2Z1B1 Canada

10 9 8 7 6 5 4 3 2 1

ISBN: 978-1-4878-1179-2

To find out more about our publications,
please visit www.royalcollins.com.

CONTENTS

INTRODUCTION

The Mariana Trench is approximately 200 km east of the Mariana Islands in the Western Pacific. Shaped like a crescent, it averages about 70 km in width and stretches for approximately 2,550 km in length. The deepest part of the trench reaches around 11 km, surpassing the Earth's highest point, Mount Everest.

Even if Mount Everest and Mount Tai were stacked together and placed in the trench, they would likely have yet to reach their deepest point. This place is irresistibly alluring for anyone who loves the sea and adventure.

Jeremy, a Swiss physicist, is also an explorer who has achieved remarkable feats in deep-sea exploration. Like Jeremy, Scott, another avid ocean adventurer, has harbored dreams of marine exploration since childhood. Jeremy and Scott met at an oceanic exhibition and planned their deep-sea journey to the Mariana Trench.

On July 11, 2048, they were on their third day in the Mariana Islands. According to the schedule, today, they were to enter the pressure chamber of the submersible to begin their deep dive. However, before that, they awaited contact from the ground base to ensure the day's plan could proceed smoothly.

After breakfast, Jeremy and Scott started their preparations. As Jeremy took out the deep-diving equipment, his smartwatch vibrated. It was a call from the ground base. Jeremy quickly set aside his equipment and pressed the receive button. The voice of the ground base commander came through: "Jeremy, Scott, your plan has been approved. We'll send the relevant materials to your smartwatches via satellite communication and update you during your dive. Maintain open communication with the base. Good luck."

Jeremy knew the base would monitor their movements if they wore their smartwatches. The satellite communication system could receive the positioning signal from the smartwatches and transmit information to the base's computer center. Unlike before, global satellite networks now form an invisible grid covering the Earth. Even in the deep sea, they could efficiently exchange information and receive support, and in case of danger, they could promptly call for help.

Jeremy and Scott exchanged excited glances, eager for the day's mission. Soon, Jeremy's smartwatch displayed a detailed plan and materials, which they projected holographically into the air. Using the satellite-scanned aerial imagery of the Mariana Islands and data from the ground base, Scott recalibrated their diving path.

At noon, Jeremy and Scott said goodbye to the surface crew and entered the submersible's pressure chamber. Eighteen minutes later, the submersible began its slow descent. After ten minutes and reaching 81 m, they stopped to adjust buoyancy and check instruments. The submersible continued its cautious descent, pausing occasionally. Beyond 200 m, water temperature and density changes lessened, and the submersible did not stop again, descending toward the seabed.

Jeremy and Scott turned off the underwater lights to observe biolumi-nescent marine life. They saw glowing plankton darting past the observation window at 690 m and again at 6,800 m. At 1:30 p.m., they reached a depth of 7,900 m. They turned on the lights but saw nothing in the bright beams.

At 9,900 m, a dull bursting sound suddenly echoed in the submersible, accompanied by a vibration. To investigate, they turned off all instruments. In the eerie silence, they heard faint scraping sounds against the hull. Rapidly, they transmitted the situation via satellite to the ground base and soon received a response. The engineers at the control center remotely accessed the submers-ible's systems through satellite communication and determined that the sub-mersible was usually descending without serious malfunction.

At 1:46 p.m., the submersible gently rested on the soft, pale-yellow seabed. Excitedly, Scott grabbed the phone to report the good news. Turning on the lights, they saw a flatfish about 30 cm long and 15 cm wide, swimming outside the acrylic observation window with slightly protruding eyes. They also spotted a red shrimp approaching the window.

After working at the front of the submersible, Jeremy opened the rear lights to look outside. Suddenly, he exclaimed, "There's a crack in the entranceway window! That's where the strange sounds were coming from." They immediately reported this unexpected situation to the surface.

Following urgent deliberation, the ground base reported back. At the tre-mendous pressures of the deep sea, the submersible's pressure chamber was enduring a total pressure of 150,000 tons, enough to compress the metal hull's diameter by 1.5 millimeters. The contraction of the metal slot at the junction of the observation window had ultimately led to a crack in the acrylic glass.

The report advised Jeremy and Scott to shorten their seabed stay from 30 to 20 minutes. Upon receiving the immediate surfacing instruction, Scott pressed the button to release the ballast pellets, and the submersible floated up like a balloon. At 4:36 p.m., Jeremy and Scott finally surfaced.

There are many regions on Earth that humanity has yet to set foot in or finds challenging to access. Yet, humans are always filled with curiosity and a desire to explore these unknown areas. Unfortunately, due to the limited coverage of ground communication, further exploration into these unknown territories is hindered. Furthermore, the limitations of ground communication prevent the development of interconnected devices, autonomous driving, the metaverse, and interstellar travel, keeping these human dreams in the realm of imagination.

Fortunately, the advent of satellite communication offers a new possibility. By extending communication networks to every corner of the Earth via satellites, whether remote deserts, vast oceans, or the vastness of space, satellite communication systems can provide stable and efficient services. This means humanity can engage in more activities in broader regions. Satellite communication is the backbone of human exploration into the unknown.

Now, let's begin with this book to formally explore the endless possibilities of satellite communication technology and the approaching era of satellite communications.

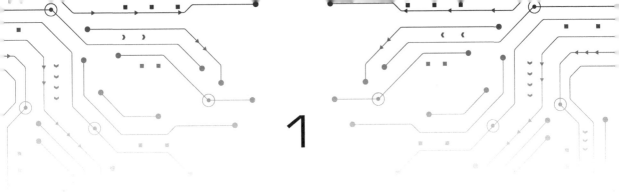

1

THE DAWN OF
THE SATELLITE ERA

1.1   The Journey of Artificial Satellites into Space

Humanity's relentless progression and ascent to civilization are rooted in an unyielding spirit of exploration. This spirit has driven us to probe our environment, venture beyond jungles into oceans, and conquer mountains and peaks. History shows that some innovations, while advancing civilization, spawn new problems. The use of oil has ignited endless wars, while plastics have led to global pollution. The future implications of lithium batteries are still uncertain. Yet, humans continually rely on innovation to forge new civilizations and rectify past destructions.

In the path of human innovation and exploration, it's clear that our ambitions extend beyond Earth. In 1957, the Soviet Union launched the first artificial satellite, marking the beginning of space exploration. Today, thousands of artificial satellites orbit overhead, and humanity has established a foothold in space.

1.1.1   The First Human-Made Satellite

Generally, an artificial satellite, simply referred to as a satellite when there's no ambiguity, is a type of spacecraft constructed by humans and the most numerous kinds in space. Satellites are crucial for modern atmospheric detection and form the backbone of Earth's communication and future interstellar communication systems. They are launched into space-by-space vehicles like rockets or shuttles and orbit Earth or other planets like natural satellites.

The history of satellites is longer than most imagine, dating back to the era of hot air balloons and sounding rockets. Their purpose: to observe Earth from the upper atmosphere. As early as AD 1500, during China's Ming Dynasty (AD 1368–1644), an official named Wan Hu (formerly known as Tao Chengdao) attempted to reach space by strapping 47 rockets to his chair and holding two giant kites. Unfortunately, the rockets exploded mid-air, and Wan Hu perished, sacrificing his life for scientific exploration. He can be considered the first "space explorer" in human history, laying the foundation for later space endeavors.

The earliest rockets appeared during China's Three Kingdoms period (AD 220–280), consisting of torches placed in arrows and launched. Gunpowder, originating from alchemy in the Spring and Autumn period, was truly invented during the Sui Dynasty (AD 581–618) and used in warfare by the late Tang Dynasty (AD 618–907). The earliest "gunpowder rockets" appeared in the Northern Song Dynasty (AD 960–1127), and rockets began to be used in warfare, eventually reaching Europe via the ancient Silk Road. Wan Hu's exploration was inspired by these inventions, though, at that time, the technological capabilities were insufficient to achieve orbit.

According to scientists, a spacecraft must reach the first cosmic velocity of 7.9 km/s and be launched horizontally to orbit Earth. If it's slower, Earth's gravity will pull it back. Once in orbit, the spacecraft will gradually slow due to atmospheric drag, eventually falling back into the dense atmosphere and burning up from friction.

When a spacecraft's speed is between 7.9 km/s and 11.2 km/s, its orbit is elliptical. The escape velocity, or the second cosmic velocity of 11.2 km/s, is needed to break free from Earth's gravity, orbit the sun, or travel to other planets.

The period between 1945 and 1955 was pivotal for the successful launch of artificial satellites, laying much of the theoretical groundwork. In October 1945, Arthur Charles Clarke published a paper titled "Extraterrestrial Relays" on *Wireless World*, proposing the feasibility of communication satellites in geostationary orbit (GEO). This paper detailed a system of geosynchronous satellites for global broadcast and television signal transmission. Clarke's predictions remarkably aligned with future developments, earning GEO the name "Clarke Orbit." His significant contributions as a science fiction writer, alongside

Isaac Asimov and Robert Heinlein, largely influenced technological innovation and exploration.

Clarke's paper discussed the principle of a geostationary satellite appearing stationary to an Earth observer and its potential to provide uninterrupted global communication services. He emphasized the importance of placing a satellite equipped with intercontinental communication devices in this orbit, where it would appear stationary relative to Earth's surface.

Clarke's idea was simple: place a satellite at a specific position above Earth, synchronizing its orbit with Earth's rotation to appear stationary to ground observers. This concept's key was continuous communication services without needing constant adjustment.

In 1945–1955, the US and the Soviet Union mainly launched experimental sounding rockets, reaching the upper atmosphere with various instruments for specific missions.

After World War II, the Soviet Union made significant advancements using German technology like the V-2 missile. The first A-4 rocket, carrying scientific instruments, was launched in 1946 to record cosmic ray flux at 112 km. The Soviets improved the A-4, leading to the V-2A rocket in 1949, reaching 212 km with an 860 kg payload.

In 1955, the US and the Soviet Union started drafting plans for artificial satellites. The Soviet lead, Sergei Korolev, published *Rocket Propulsion* in 1931 and designed the first liquid-fueled rocket in 1932. Despite political imprisonment, Korolev continued his work, eventually overseeing the design of the first Soviet missiles and initiating satellite launch plans.

In 1956, learning of the US "Vanguard" satellite, Korolev proposed launching two simple satellites using modified intercontinental missiles, which was approved. On August 3, 1957, his intercontinental missile test succeeded, overcoming technical challenges in rocket bundling and separation.

Finally, on October 4, 1957, using a bundled rocket, Korolev successfully launched "Sputnik 1" from the Baikonur Cosmodrome, heralding a new era of outer space exploration with artificial satellites.

1.1.2   The Advent of "Explorer 1"

On October 4, 1957, the launch of "Sputnik 1" by the Soviet Union's R7 Intercontinental Ballistic Missile from the Baikonur Cosmodrome marked the tangible beginning of the space age. This first artificial satellite, made of chrome-plated alloy, weighed 83.6 kg. Shaped like a sphere with a diameter of 584 cm, it orbited

Earth every 96 minutes, with an apogee of 986.96 km and a perigee of 230.09 km. After 92 days in orbit, traveling approximately 1,400 times around the Earth and covering a distance of 60 million km, "Sputnik 1" sent simple signals to Earth through its four external antennae. Radio enthusiasts worldwide received these signals from space. On January 4, 1958, after its successful mission, "Sputnik 1" burned up upon re-entering Earth's atmosphere.

Though humanity's first satellite did not last long in space, its launch had a profound impact. Journalists worldwide described it as a "cosmic shock," a significant scientific achievement, and one of the most extraordinary events in world history. The US media dubbed the satellite the "Red Moon." Many tried to spot "Sputnik" in the twilight sky, seeing only the rocket's central part until it burned up, not the tiny sphere less than 1 m in diameter. The US author Ray Bradbury reflected, "The night Sputnik 1 crossed the sky for the first time, I looked up and pondered our future. That tiny star swiftly moving from one side of the sky to another represented humanity's future. Despite the Russians' impressive effort, I knew we would soon follow them and claim our rightful place in the sky."

Furthering their achievements, the Soviet Union successfully launched "Sputnik 2" on November 3, 1957. It orbited Earth in an elliptical path with an inclination of 65.33°, apogee of 212 km, and perigee of 1,660 km. Notably, "Sputnik 2" carried a female dog named Laika, the first living being to orbit Earth, providing valuable data on the biological effects of spaceflight.

The launch of "Sputnik 1" and "Sputnik 2" left Americans surprised and somewhat embarrassed, as they had not yet successfully launched their satellite. However, this quickly changed on January 31, 1958, with the successful launch of "Explorer 1" by the US.

Launched from Cape Canaveral by a Jupiter-C rocket, "Explorer 1" was designed like a pencil, allowing it to spin like a bullet while orbiting Earth. This spin provided stability in orbit, becoming a fundamental technique for satellite stabilization. During its mission, "Explorer 1" made a significant discovery: a radiation belt encircling Earth within its magnetic field.

Following this success, the US launched "Vanguard 1" on March 17, 1958. This metal sphere, weighing only 3 pounds (1.5 kg) and measuring 6.5 inches (16.5 cm) across, was smaller than a basketball. Soviet leader Nikita Khrushchev mockingly called it the "grapefruit satellite." Despite its size, "Vanguard 1" was groundbreaking, being the first satellite to use solar cells for power, drawing energy from six matchbox-sized panels.

As part of the 1957–1958 International Geophysical Year scientific program, the US Naval Research Laboratory was tasked with building this satellite. The Vanguard team was a precursor to NASA. When NASA was established in 1958, many naval researchers joined and formed the core of the Goddard Space Flight Center. When the project transitioned from the Navy to NASA, the researchers continued with it. They designed the Vanguard Project as an experiment to study how the space environment affected satellites. Once in orbit, researchers meticulously tracked its trajectory, observed how it deviated from predictions, and used this information to refine their understanding of the Earth's upper atmosphere.

1.1.3   China's Voice in Space

Following the successful satellite launches by the Soviet Union and the US, 1970 marked the year when space first echoed with a voice from China. That year, China's first artificial satellite, "Dong Fang Hong 1" (The East Is Red no. 1), was successfully launched, heralding a new era in Chinese space history. Weighing 173 kg and designed as a 1 m diameter polyhedron sphere with four whip-like 3 m-long antennae, the satellite orbited Earth for 114 minutes. While in orbit, it broadcasted the "The East Is Red" melody at 20.002 MHz, simultaneously transmitting telemetry parameters and scientific detection data. Its launch date, April 24, is now celebrated annually as China's Space Day.

The successful launch of "Dong Fang Hong 1" was a remarkable achievement for China then. Although the Soviet Union and the US had launched satellites as early as October 1957 and January 1958, China was far from ready regarding economy, talent, and modern scientific research. Even in 1959, the satellites and rocket models displayed to the public were handcrafted by young scholars.

In early 1958, three renowned scientists, Qian Xuesen, Zhu Kezhen, and Zhao Jiuzhang, proposed the development of China's artificial satellite, garnering high-level attention. On May 17th, Mao Zedong announced at the Second Plenary Session of the 8th Congress of the Communist Party of China: "We also need to make artificial satellites." Subsequently, the Chinese Academy of Sciences prioritized the development of an artificial Earth satellite named the 581 Project. In July, a leadership group for the 581 Project was formed, led by Qian Xuesen with Zhao Jiuzhang and Wei Yiqing as deputies. However, due to economic and technical constraints, the satellite project was adjusted to prioritize the "Two Bombs" (nuclear and hydrogen bombs) project, shifting the focus to sounding rocket research.

By 1964, after overcoming three years of economic hardship, China's national economy began to improve, and the successful test launch of its medium-range surface-to-surface missile laid the groundwork for resuming the satellite project. In December 1964, Zhao Jiuzhang wrote to Premier Zhou Enlai, suggesting including the satellite project in the national plan. In January 1965, Qian Xuesen submitted a proposal to Marshal Nie Rongzhen to resume the satellite program. On August 2, 1965, Premier Zhou Enlai approved the proposal at a special committee meeting, thus reigniting China's satellite project, now named the 651 Plan.

In October 1965, a national plan validation meeting for the "651" project lasted 42 days at the Beijing Friendship Hotel. After intense discussions, the basic plan for "Dong Fang Hong 1" was established: "Reach Orbit, Maintain Tracking, Be Audible, and Be Visible." This seemingly simple word phrase represented a complex modern scientific system.

"Reach Orbit" meant that the satellite had to enter its predetermined orbit, requiring extensive calculations—done manually with hand-cranked calculators, taking a whole year to compute a single orbit.

"Maintain Tracking" refers to the satellite's tracking system's effectiveness and ability to confirm its trajectory. For this, "Dong Fang Hong 1" first adopted Doppler tracking technology and a transponder as a double assurance to understand the satellite's flight path.

"Be Audible" was about broadcasting the distinctly Chinese melody "The East is Red" into space.

"Be Visible" involved making the satellite observable to the naked eye of ordinary people. To achieve this, Shen Zuwei, a satellite's general assembly team member at a research institute under the Ministry of the Seventh Machine-Building Industry, spent nearly a year developing an observation skirt for the final stage rocket. This skirt, made of aluminized polyimide satin attached to the satellite's 72 surfaces, reflected sunlight with sufficient brightness while withstanding space temperature variations and radiation.

Finally, on April 24, 1970, the "Dong Fang Hong 1" satellite, carried by the Long March 1 rocket, successfully entered its predetermined orbit, completing a mission of over twenty days and ceasing signal transmission on May 14. The successful launch of "Dong Fang Hong 1" marked China as the fifth country to independently launch an artificial satellite, following the Soviet Union, the US, France, and Japan. This achievement was a breakthrough in China's multi-stage rocket technology, laying a solid foundation for successfully launching "Dong Fang Hong 1" and creating favorable conditions for future endeavors. China completed the project in less than five years, from its official inception to its successful launch.

Since the launch of "Sputnik 1" in 1957, humanity has sent thousands of satellites into space. These satellites have changed our methods of managing Earth and altered modern warfare strategies, enabling modern technological exploration beyond Earth's atmosphere and revealing more possibilities for the future. Today, space exploration is no longer a choice but an imperative.

1.2 How Many Astronomical Satellites Are There?

Since the launch of humanity's first artificial satellite, "Sputnik 1," in 1957, we have entered the era of artificial Earth satellites. Before this, for over 4.5 billion years, Earth had only one natural satellite, the moon. "Sputnik 1" became another satellite bestowed by intelligent life upon Mother Earth, marking the grand opening of human space history. However, early satellites primarily validated space technology, with basic functions limited to broadcasting and simple scientific experiments. Rapidly, the potential applications of artificial satellites were realized. Over the past 65 years of space exploration, the satellite family has expanded significantly with various classification methods. The most common methods categorize satellites based on their orbital altitude, orbit path, and specific uses.

1.2.1 Satellites at Different Orbital Altitudes

Satellites are categorized into Low Earth Orbit (LEO), Medium Earth Orbit (MEO), and High Earth Orbit (HEO) based on their orbital altitudes, each having unique characteristics and purposes.

1.2.1.1 LEO Satellites

LEO satellites, also known as near-Earth satellites, operate in orbits close to the Earth's surface, typically below 2,000 km in altitude. Due to their proximity to Earth, LEO satellites offer shorter transmission delays and reduced path loss, providing end users with efficient and cost-effective data distribution services. A single LEO satellite typically covers an area of several hundred to a thousand kilometers in diameter. When numerous satellites form a constellation at this altitude, they can achieve seamless global coverage. The communication transmission delay between the ground and these satellites is only a few milliseconds, meeting the real-time requirements of applications like autonomous driving and drone remote control. The most representative LEO satellite communication systems currently include Starlink, Iridium, and Globalstar.

1.2.1.2 MEO Satellites

MEO satellites operate between 2,000 km and 36,000 km, serving as an extension and supplement to terrestrial Internet systems and facilitating global personal mobile communications. They also function in satellite navigation systems. MEO satellites combine the advantages of HEO and LEO satellites, achieving global communication coverage and effective frequency reuse. In addition to deploying numerous satellites, MEO systems involve complex inter-satellite networking and control switching.

Representative MEO satellites include the Inmarsat-P, Odyssey, MAGSS-14, and the BeiDou Navigation Satellites System.

The Inmarsat-P system (the third generation of the International Maritime Satellite) is a new personal satellite mobile communication system, also known as the Project-21 plan. The first and second generations relied solely on geostationary satellites, covering only regions above 70° latitude. The third generation combines synchronous and medium orbit satellites. Its four synchronous satellites, equipped with over 10 m antennae and onboard processing capabilities, facilitate handheld communication. The MEO component comprises 12 satellites at about 10,000 km altitude, featuring onboard processing. The system aims to provide 90% visibility of satellites under various conditions in the service area, with handheld devices transmitting at 0.25–0.4 W, ensuring 95% communication reliability.

The Odyssey system, built by TRW, consists of 12 MEO satellites distributed over three orbital planes with a 55° inclination, utilizing L, S, and Ka frequency bands. Each satellite has 19 beams and a total capacity of 2,800 circuits, each serving 100 users. Thus, the system can serve 2.8 million users globally. The construction cost is approximately $2.7 billion, and the satellites have a designed lifespan of 12–15 years.

The MAGSS-14 system, developed by the European Space Agency, comprises 14 satellites at 10,354 km altitude, distributed over seven orbital planes with a 28.5° inclination. At this altitude, satellites complete an orbit in a quarter of a sidereal day (23 hours and 56 minutes). With a user elevation angle of 28.5°, the maximum slant range is 12,500 km, yielding a satellite coverage radius of 4,650 km. Each satellite's 37 beams can cover the globe.

China's BeiDou MEO satellites operate at approximately 21,500 km with a 55° inclination. Orbiting the Earth, they form a network to provide global satellite positioning signals. The BeiDou MEO constellation has a regression feature of 7 days and 13 orbits.

1.2.1.3 HEO Satellites

HEO satellites, including GEO and Inclined Geosynchronous Orbit (IGSO) satellites, operate at altitudes of 36,000 km. GEO satellites have a 0° orbital inclination, while IGSO satellites have an incline greater than 0°. Both types have orbital periods strictly synchronized with Earth's rotation, appearing "stationary" relative to the ground and covering fixed regions.

HEO satellite communication systems offer simple frequency coordination, longer operational lifespans, and lower initial construction costs. Theoretically, a single HEO satellite can cover an entire hemisphere, forming a regional communication system. This system provides mobile communication access services for any location within its coverage area. With just three HEO satellites, global coverage is possible. However, damage to a single satellite can potentially disrupt the entire communication system.

Representative HEO satellites include the North American Satellite Mobile Communications System and the INMARSAT.

The North American Satellite Mobile Communications System, the world's first regional satellite mobile communication system, primarily provides public wireless and dedicated communication services. Ground gate stations connect to terrestrial telephone networks via wired lines. Once the network control center allocates radio frequency channels to the gate stations, communication channels are established between mobile and fixed users.

The INMARSAT, the earliest HEO satellite mobile system, was initially a military satellite communication system using Marisat satellites by Communications Satellite Corporation. In the mid-1970s, part of its services was extended to ocean-going vessels for enhanced safety. In 1982, the system was managed by the International Maritime Satellite Organization, providing global maritime satellite communication services. In 1985, the convention was amended to include aeronautical communication, and in 1989, the service expanded from maritime to terrestrial. Today, it is an international satellite mobile communication organization with 72 member countries, controlling numerous voice and data systems in 135 countries. The Ministry of Transport and the China

Transport Telecommunications & Information Center represent China in this organization.

1.2.2 Satellites in Different Orbits

Satellites are categorized based on their orbits, such as GSO satellites, Sun-Synchronous Orbit (SSO) satellites, and Polar Orbit satellites.

Geosynchronous Earth Orbit satellites operate in prograde orbits with periods matching Earth's rotation. In these orbits, satellites maintain a constant velocity, aligning with Earth's rotation and appearing stationary relative to a specific point on Earth. This orbit is particularly advantageous for communication and weather satellites. Communication satellites can provide stable services in specific areas, while weather satellites enable meteorological monitoring and forecasting.

SSO satellites rotate around Earth's axis in the same direction as Earth's revolution, with a rotation speed equal to the average angular velocity of Earth's orbit. Typically, these orbits do not exceed 6,000 km in altitude. The main feature of SSO is that satellites pass over the same point at consistent times, facilitating Earth's surface observation under uniform lighting conditions. This consistency in data acquisition under identical lighting conditions at different times is beneficial for scientific research and Earth observation.

Polar Orbit satellites, with an inclination of 90°, pass over the Earth's poles in each orbit, allowing them to oversee the entire Earth's surface. This orbit is characterized by its ability to provide global coverage, making it suitable for global monitoring tasks. Weather satellites, Earth resource satellites, and reconnaissance satellites often use this orbit. Polar Orbit satellites can gather information globally, meeting the needs of monitoring different regions and time points on Earth.

1.2.3 Satellites for Different Purposes

Satellites are divided into three categories based on their functions: communication, navigation, and remote sensing.

1.2.3.1 Communication Satellites

As the name suggests, communication satellites are primarily used for communication. In this process, they serve as relay stations for radio communication, facilitating communication between satellites and ground stations or spacecraft.

Communication satellites are modern society's beacons and swift horses for information transfer, efficiently and accurately transmitting information between users. The types of information they handle are vast, ranging from television and live data, mobile communication data, broadcast data, and Internet data to scientific data from lunar landers/rovers and Mars missions. Consequently, communication satellites are the earliest, most deeply researched, and most diverse type of satellite.

Communication satellites are further categorized into four types: direct-broadcast satellites for public television and radio transmission; maritime communication satellites for communication between sea, air, and land, also serving rescue and navigation tasks; tracking and data relay satellites for control and data transmission between spacecraft and ground stations; and navigation satellites, guiding planes, ships, vehicles, and other entities along chosen routes.

For instance, due to its vast high-latitude territories, Russia launched unique Molniya orbit communication satellites, enabling extended coverage over the northern hemisphere. Recent years have seen a surge in LEO communication satellite networks, with SpaceX's Starlink project being the most prominent, planning to deploy over 40,000 satellites for global Internet service. China's Queqiao Satellite launched to the moon, and the Tianwen-1 Mission Orbiter to Mars also functioned as signal relays.

1.2.3.2 Navigation Satellites

Navigation satellites primarily provide positioning services for ground, sea, air, and space users and possess communication properties. Unlike the two-way user-satellite-user information transfer model of communication satellites, navigation satellites typically employ a satellite-to-user one-way information transfer method. They continuously broadcast their precise location and time data, allowing users to calculate distances to multiple satellites and thus determine three-dimensional coordinates and time information.

Satellite navigation systems offer the advantages of traditional navigation systems, such as all-weather global high-precision passive positioning. Time-based satellite navigation systems resist interference and provide continuous three-dimensional coverage and high-precision positioning for global and near-Earth space.

The four major global satellite navigation systems operating are the US GPS, Europe's Galileo, China's BeiDou, and Russia's GLONASS.

The GPS, developed jointly by the US Army, Navy, and Air Force in the 1970s, aims to provide real-time, all-weather, and global navigation services for land, sea, and air domains. It also serves military and non-military purposes, such as intelligence gathering and emergency communication. After 20 years of experimental research, the GPS constellation of 24 satellites was completed in March 1994, costing about $30 billion, with a global coverage rate of 98%. GPS technology has evolved into a multi-domain, multi-mode, and multi-purpose international high-tech industry.

Applications of GPS include land-based uses such as vehicle navigation, atmospheric physics observation, geophysical resource exploration, engineering surveying, deformation monitoring, crustal movement monitoring, and municipal planning control; marine applications like ocean-going ship route determination, real-time ship scheduling and navigation, maritime rescue, treasure hunting, hydrogeological surveying, and offshore platform positioning; and aerospace applications such as aircraft navigation, remote sensing attitude control, low-orbit satellite orbit determination, missile guidance, aerial rescue, and manned spacecraft protection.

The Galileo navigation system, comprising 30 satellites and offering positioning accuracy within 1 m, is primarily for civilian use. The European Galileo system began operating in 2014, with its first test satellite successfully launched in 2005 and positioning services initiated in 2008. On January 24, 2018, the EU announced the relocation of the Galileo satellite navigation system's security monitoring center from the UK to Spain. Galileo, named after the 17th-century Italian astronomer, started official operations in 2016 and completed its full satellite network by 2020, providing global navigation services.

China's BeiDou Satellite Navigation System, independently developed by China, is the third mature satellite navigation system after GPS and GLONASS. From the inception of the BeiDou-1 project in 1994 to its completion in June 2020, the BeiDou system has evolved over 26 years. It is the only system that includes MEO and Geosynchronous Earth Orbit (inclined + stationary), offering unique two-way short message services. BeiDou, GPS, GLONASS, and Galileo are recognized by the United Nations Satellite Navigation Committee.

GLONASS, developed by the Soviet Union in 1982 and inherited by Russia after the Soviet Union's dissolution, briefly lost most of its satellites and functionality in the early post-Soviet era. GLONASS has slight technical differences compared to the US GPS, European Union Galileo, and Chinese BeiDou systems. The GLONASS constellation consists of 21 operational satellites and three spares, evenly distributed across three near-circular orbital planes, each 120° apart. Each plane contains eight satellites spaced 45° apart at an altitude of 23,600 km, an orbital period of 11 hours and 15 minutes, and an inclination of 64.8°. Operational since 2007, initially providing positioning and navigation services within Russia, GLONASS expanded globally by 2009. The system's primary services include determining coordinates and movement speeds of targets on land, at sea, and in the air. Russia is currently modernizing the GLONASS system.

1.2.3.3 Remote Sensing Satellites

Remote sensing satellites are those used as platforms for outer space remote sensing. The technology employing satellites as remote sensing platforms is known as satellite remote sensing technology. These satellites can cover specified areas within a set time. Remote sensing satellites can continuously monitor designated regions when operating in GSO. Remote sensing satellites require ground station systems, through which they can relay data relevant to agriculture, forestry, oceanography, land management, environmental protection, and meteorology.

Meteorological remote sensing satellites are primarily used to monitor atmospheric, terrestrial, and oceanic conditions globally. Based on the remote sensing data acquired, weather maps can be drawn, cyclones, typhoons, hurricanes can be detected, and cloud top and surface temperatures can be determined, enabling timely warnings for extreme weather events. Currently, the primary remote sensors used in meteorological satellites are Visible Light-Infrared Scanning Radiometers and High-Resolution Scanning Radiometers. Additionally, meteorological satellite remote sensing data are extensively applied in non-meteorological activities like navigation, fishing, and monitoring crop growth.

Land resource satellites are designed to assist in resource mapping and surveying by obtaining Earth's surface image data with onboard remote sensors. Typically operating in circular sun-synchronous orbits, these satellites have observation cycles of ten to thirty days. Remote sensing data from land resource satellites are widely applied in numerous fields, including national land surveys, geological investigations, resource exploration, agricultural and forestry surveys and planning, engineering site selection, and coastal topographic mapping. They provide rapid data acquisition and technical support for governments and businesses to formulate sound policies and plans.

Ocean resource satellites are tasked with detecting surface conditions and monitoring ocean dynamics. These satellites are generally equipped with synthetic aperture radars, radar altimeters, microwave radiometers, and infrared radiometers. Ocean resource satellites provide continuous, comprehensive, and synchronized global ocean observation data, such as ocean wave height, length and spectrum, ocean wind speed and direction, ocean temperature, currents, circulation, sea state, and global sea level. The advent and application of ocean resource satellite technology have significantly advanced oceanographic research, giving rise to the new field of satellite oceanography.

1.3 Satellite Ascendancy: Soaring High in Space Economics

In an era marked by the rapid development of space economics, the satellite industry, as a crucial component, is increasingly becoming a vital force in advancing scientific and technological progress and serving economic and social development. It also carries the key technology for humanity's venture beyond Earth and into the far reaches of space.

1.3.1 Expanding Market Size

A complete satellite system consists of a space system and a ground system, both functioning in a complementary and sustained manner.

Regarding space systems, although satellites vary widely in their application fields, they generally comprise two main subsystems: the payload and the support system. The payload directly accomplishes specific space missions, while the support system ensures the normal operation of all subsystems on the satellite from launch to the end of its working life. Common components of the support system include structural, thermal control, power, attitude and orbit control, and telemetry, tracking, and command systems.

The ground system mainly consists of ground control and application systems. The ground control system includes tracking and measurement, telemetry, telecommand, real-time computer processing, display recording, time synchronization, communication, and post-mission data processing subsystems. It facilitates mobile user access to the core network and controls and manages the space segment's operations and user services. The satellite ground application system varies according to the satellite's application field but mainly comprises various user terminals, including handheld, portable, embedded, vehicular, shipborne, and airborne terminals.

The US Satellite Industry Association (SIA) categorizes the satellite industry into four segments: satellite launch, manufacturing, services, and ground equipment.

According to the SIA report, in 2021, global revenues for satellite launch, manufacturing, services, and ground equipment manufacturing were $5.7 billion (+8% YoY), $13.7 billion (+12% YoY), $118 billion (+0.2% YoY), and $142 billion (+5% YoY), respectively. The commercial force in satellite launch is rising, with the commercial satellite industry launching 1,713 satellites into orbit in 2021, a more than 40% increase YoY. Moreover, with advancements in satellite manufacturing, launching technologies, and large-scale deployments of LEO satellites, the satellite manufacturing industry is benefiting from the surging demand. Regarding services, while satellite TV broadcast and radio growth have been under pressure, factors such as remote work are catalyzing growth. Future growth in satellite services will further manifest in satellite broadband, mobile communication, and remote sensing.

Meanwhile, according to the Satellite Industry Association's State of the Satellite Industry Report, global aerospace economic revenue grew from $322.7 billion in 2014 to $386 billion in 2021, maintaining steady single-digit growth. The global satellite industry size was $246 billion in 2014, growing at over 2% annually until 2019. After a growth hiatus in 2019 and 2020 due to the pandemic and industry restructuring, it achieved a 3% YoY growth in 2021, reaching $279 billion. The global satellite industry continues to expand, consistently accounting for over 70% of total aerospace economic revenue.

1.3.2 Highlighting Strategic Value

As a core component of the space economy, the strategic value of the satellite industry is increasingly prominent amid intense global competition for space resources.

Satellites in operation require specific orbital positions in outer space, and the number of satellites that near-Earth orbit can accommodate is limited. Additionally, the frequency bands with low loss, crucial for developing space services, are finite. This indicates that orbital and spectrum resources are scarce and exclusive strategic resources in space.

According to a report by the Chinese research institute CCID Consulting, near-Earth orbit can accommodate about 60,000 satellites. By 2029, it is expected that nearly 57,000 satellites will be deployed in near-Earth orbit, indicating a high level of congestion. Also, the Ku and Ka communication bands mainly used by low-orbit satellites are gradually becoming saturated. As per the International Telecommunication Union's Radio Regulations, near-Earth orbit and frequencies are allocated on a "first-come, first-served" basis, highlighting the scarcity of orbital and spectrum resources and the ongoing rush by countries to secure these strategic assets in space for a competitive advantage.

In recent years, China's policies have continuously propelled the in-depth development of the satellite industry. In November 2016, the 13th Five-Year Plan for National Strategic Emerging Industries Development aimed to strengthen and expand the satellite and application industry. The goal was to build an autonomous, open, safe, and reliable national civil space infrastructure, accelerate the integration of satellite applications with infrastructure, and by 2020, have a fully functional national civil space infrastructure to meet major business needs, achieve autonomous space information application, and form a comprehensive satellite and application industry chain. During the 13th Five-Year Plan period, China accelerated satellite launches, deploying 255 satellites.

In March 2021, the 14th Five-Year Plan for National Economic and Social Development and the Long-Term Goals for 2035 was released, calling for the construction of a global coverage, efficient operation communication, navigation, and remote sensing space infrastructure system, including the construction of commercial space launch sites.

In November 2021, the 14th Five-Year Plan for Information and Communication Industry Development was released, emphasizing strengthening the top-level design and overall layout of satellite communication, promoting the coordinated development of high-orbit and medium-low-orbit satellites, and advancing the deep integration of satellite communication systems with terrestrial information and communication systems. The plan aimed to initially form a global information network integrating heaven and earth, providing global information network services for various land, sea, air, and space users. It encouraged participation in the formulation of international standards for satellite communication, innovation in satellite communication applications, and large-scale application of the BeiDou Satellite Navigation System in the information and communication field, promoting its application in aviation, maritime, public safety, emergency, transportation, energy, and other fields.

In December 2021, the 14th Five-Year Plan for Digital Economy Development was released, calling for the active and prudent promotion of the evolution and upgrade of space information infrastructure, accelerating the layout of satellite communication networks, and enhancing the supporting capabilities of satellite communication, remote sensing, and satellite navigation positioning systems to build a global coverage, efficient operation communication, remote sensing, and navigation space infrastructure system.

In December 2021, the 14th Five-Year Plan for National Emergency System was released, requiring the improvement of emergency satellite observation constellations, constructing an integrated space, sky, land, and sea coverage disaster and accident monitoring and early warning network; steadily advancing the construction of satellite remote sensing networks; developing integrated emergency disaster reduction satellite application systems and autonomous operation management platforms; promoting the application of space-based satellite remote sensing networks in disaster prevention, reduction, rescue, emergency rescue management; constructing a satellite communication management system based on Tiantong, BeiDou, satellite Internet, and other technologies to achieve unified dispatch and comprehensive application of emergency communication satellite resources; building a high-throughput satellite emergency management system, expanding and reconstructing the satellite emergency management comprehensive service system.

It is foreseeable that during the 14th Five-Year Plan period, China will continue to improve the space infrastructure system, deepen application innovation, and enhance the satellite industry's ability to serve the real economy.

1.3.3 Technological Acceleration in Development

In recent years, satellite development has evolved toward miniaturization and standardization, optimizing costs and laying the foundation for mass satellite launches and networking. Small satellites have gained wide usage and attention from various countries. Concurrently, rocket launches have achieved multiple satellites per launch, and China is actively developing reusable rockets, expecting further control over launch costs to support frequent satellite launches.

Looking at the miniaturization and standardization of satellite manufacturing, satellites weighing less than 1,000 kg are collectively termed "small satellites." Applying microelectronics, micromechanics, nanotechnology, and integrated circuit manufacturing has enabled small satellites to have short development cycles, low costs, lightweight, small size, flexible launch methods, and the ability to form "virtual large satellites" in constellations. At the same time, they achieve high performance and wide application across multiple fields.

Small satellites are used in communication, remote sensing, scientific research, military, and interplanetary exploration and are increasingly valued by various countries. Under such circumstances, small satellite manufacturing technology has continually advanced, establishing the foundation for mass production launch and networking. Since the 21st century, the US has continuously experimented with military small satellite constellations, undertaking projects like "Blackjack." SpaceX's Starlink program has launched numerous small satellites to establish a global satellite Internet. As of October 1, 2023, SpaceX has launched 5,200 Starlink satellites, with 4,849 in orbit, 4,797 operational, and 4,199 in active service. The UK's OneWeb plans to launch 600 small satellites to build a high-speed global telecom network. China is also planning to launch 12,000 satellites to secure prime space positions and frequencies.

Recognizing the scarcity of orbital resources, particularly in near-Earth orbits, has accelerated the development of the satellite industry. The number of small satellites launched globally is rapidly growing, becoming a significant component of satellite launches, while the launch numbers for medium and large satellites remain stable. Using 1,000 kg as the dividing line between small and medium-large satellites, according to the UCS Satellite Database, 1,697 satellites were launched globally in 2021. Excluding satellites of unknown weight, there were 1,629 small satellites and 42 medium-large satellites. Since 2016, the annual compound growth rate of small satellite launches is about 94%, while the number of medium-large satellite launches has remained stable. The proportion of small satellites in global launches has rapidly increased, from about 53% in 2015 to 96% in 2021.

Simultaneously, the average weight of globally launched satellites continues to decline. According to the UCS Satellite Database, the average weight of satellites launched globally in 2021 was approximately 286.8 kg, only about one-sixth of that in 2016. Enhanced rocket carrying capacity and advancements in recovery technology have facilitated dense, short-cycle satellite launches. SpaceX, with its reusable Falcon 9 rockets and multi-satellite launch technology, has reduced launch costs and achieved dense Starlink satellite launches.

China is also actively developing reusable rockets, expecting further cost reductions. The upgraded version of the Long March 8 rocket, CZ-8R, is China's first targeted recovery rocket in development, with its first recovery planned for 2025. In February 2022, China Aerospace Science and Technology Corporation (CASC) introduced a new vertical takeoff and landing recovery scheme for a manned launch vehicle at an international seminar. Moreover, China's breakthroughs in multi-satellite launch technology are promising. In 2015, the Long March 6 rocket achieved a 20-satellite launch; in 2022, the Long March 8 Yao-2 rocket set a new record with 22 satellites in a single launch using a shared "ride-sharing" mode. These advancements are expected to impact China's aerospace technology development profoundly.

1.4   The Dawn of the Satellite Era

As a bridge connecting Earth to space exploration, the satellite industry began its expansive journey by launching the first artificial satellite. After its initial developmental stages, the global satellite industry now stands in a new era marked by rapid growth and profound transformation. With satellites increasingly integral in social production and daily life and more capital entering the satellite industry, we are entering an unprecedented satellite era.

1.4.1   Ubiquitous Satellite Applications

Today, satellites have become an omnipresent element in our daily lives. Serving purposes ranging from meteorological and astronomical observation to military and space technology research, this pivotal technology, developed over 70 years, has become indispensable to human society.

For instance, in meteorological and astronomical observation, satellites monitor large-area weather changes and provide precise forecasts for smaller regions, allowing people to be better prepared before every journey. They also facilitate real-time monitoring of natural disasters like typhoons, floods, and forest fires and provide intelligence on fishing grounds, mountainous regions, or mineral resources. Satellite observations enhance the efficiency of natural resource exploitation and disaster relief efforts.

Regarding information transmission, satellite communication involves transferring text, voice, images, videos, and data. Common civilian services include satellite phones, television broadcasting, and international messaging. Satellite communication is characterized by its extensive coverage but limited bandwidth, which makes it suitable for addressing issues in remote areas. Current systems like Inmarsat and Iridium fit this category. SpaceX's Starlink aims to provide high-speed Internet services in remote areas with hundreds of megabits per second. Another type of high-throughput satellite communication primarily serves fixed terrestrial services, replacing fiber optics, especially for long-distance information transmission, which relies on undersea cables. In the future, 5G and the Internet of Things (IoT) will also be significant arenas for satellite communication.

In navigation and positioning, the satellite navigation and location services industry, integrating information, manufacturing, and service industries, is strategically emerging with tremendous development prospects. Considered the next global information growth point following the Internet industry, satellite navigation and positioning services have become essential features of smartphones and other terminal devices. With the formation of the mobile Internet ecosystem, location-based services are bringing transformative changes to the Internet, wireless communication, the IoT, and vehicle networking, significantly altering our lifestyles. Embedding navigation modules in smartphones, wearable devices, and autonomous vehicles allows people to access geographic location information anytime, anywhere. The intersection of navigation systems with the Internet, artificial intelligence (AI), 5G, and big data drives the functional upgrade of various terminal devices, enhancing user experiences. For instance, high-precision positioning enables smart transportation and logistics services like bike-sharing, ride-hailing, and online shopping. Future developments include automated driving assistance, AR virtual tourism, and more. Currently, 80% of social life content is related to location services, with check-ins at scenic spots, nearby dining information queries, route navigation, real-time public transportation updates, and drone aerial photography. "Location" and "navigation" have become frequent terms in daily life, making satellite navigation and location services indispensable in intelligent living.

From micro to macro, from individual to the entire world, the impact of satellite applications is everywhere, quietly yet significantly driving the advancement of human technology and living standards. The development of satellite technology has fundamentally changed people's lives and will continue to influence our way of life.

1.4.2　Satellites Launching into the Future

LEO satellite Internet is becoming an essential tool for realizing smart aviation, marine, and agriculture and a vital platform for countries to conduct major scientific research. However, orbital resources are a precious, non-renewable commodity, a global consensus. This has intensified the competition for space in LEO, with countries and commercial enterprises vying for these valuable resources.

Elon Musk's Starlink project is arguably the most ambitious satellite Internet plan in the world today. SpaceX proposed the Starlink project as a LEO satellite Internet constellation system composed of several satellite constellations and ground stations. Once completed, the system will provide high-speed Internet access services globally with its constellation of 42,000 LEO satellites. In 2022 alone, 2,022 communication satellites were launched globally, with 1,721 belonging to the Starlink project. By the end of 2023, the number of communication satellites launched as part of the Starlink project exceeded 5,000.

Beyond SpaceX's Starlink, several other players have joined the satellite networking wave. Following Starlink, the UK's OneWeb has the world's second-largest satellite constellation. In March 2023, NewSpace India Limited successfully launched OneWeb's first-generation constellation of 36 satellites, bringing the operational constellation to 618 satellites and nearly achieving global service capability. OneWeb's Internet access services have expanded to cover most of Europe, including Austria, Italy, France, Portugal, and most of the US.

Amazon's Kuiper Project also plans to launch 3,236 near-Earth orbit satellites, investing over $10 billion. In February 2023, the Federal Communications Commission (FCC) approved Amazon's orbital debris mitigation plan, authorizing the company to deploy its Kuiper Ka-band LEO broadband satellite constellation and provide operational services. On April 5, 2023, Amazon announced agreements with three commercial space companies for 83 launches: 38 Vulcan Centaur launches with ULA, 18 Ariane 6 launches with Arianespace, and 12 New Glenn launches with Blue Origin, with the option to purchase 15 additional launches from Blue Origin.

On August 12, 2023, TELESAT announced a contract with MDA, a Canadian aerospace technology company, to build 198 advanced satellites for the Telesat Lightspeed LEO constellation project. The first launch of the Telesat Lightspeed satellites is planned for mid-2026, with global service set to commence once the 156th satellite is in orbit (expected by the end of 2027).

In the Chinese market, satellite launches are also accelerating. In April 2021, two remote sensing satellites, "Qilu-1" and "Qilu-4," owned by the Shandong Institutes of Industrial Technology, were successfully launched into orbit by the "Long March 6" rocket, marking the first commercial LEO satellite launch of the year. Joining them were several other satellites, including the "Foshan-1" optical remote sensing satellite developed by Beijing Gengyu Muxing Space Technology Co., Ltd. for the Guangdong Ji Hua Laboratory, the "Taijing-2-1" optical remote sensing satellite developed by Beijing Minospace Technology Co., Ltd, two remote sensing satellites "Golden Bauhinia 1" and "Golden Bauhinia 1-02" developed by ZEROG Lab, as well as "Tianqi 9," and Origin Space's NEO-1.

Subsequently, a series of commercial remote sensing satellite projects like "Beijing-3" developed by Twenty-First Century Space Investment and China Spacesat Co., Ltd., "Haisi-2" for nearshore shallow marine ecological environment observation by Xiamen University, "Yangwang-1" for asteroid resource exploration by Origin Space, along with four "Jilin-1" satellites were launched in June and July of 2021.

In October, China Spacesat Co., Ltd.'s orbital atmospheric density detection satellite, commercial meteorological detection satellite, Nanjing University of Science and Technology's "Tianyuan-1" satellite, USPACE Technology Group Limited and COMMSAT's jointly developed "Golden Bauhinia 2," Gengyu Muxing's low orbit navigation enhancement test satellite, and China HEAD Aerospace Technology Co.'s two "HEAD-2" satellites along with Shanghai Lizheng Satellite Application Technology Co., Ltd.'s VDES traffic test satellite were successfully launched. This marked the 37th launch of the year. In December, Tianjin University No. 1 meteorological observation satellite developed jointly by Changguang Satellite and Tianjin Yunyao Aerospace, along with Lize-1, Baoyun, Golden Bauhinia 5, Golden Bauhinia 1, carried by the private rocket company Galactic Energy, successfully reached orbit. By the end of 2021, China had launched 97 satellites.

However, China's "SkyNet" constellation has not yet been widely deployed. As the launch of LEO satellites accelerates and countries and commercial companies race to secure these resources, the space available for latecomers is increasingly limited, making orbital resources a critical threshold. Recognizing the importance of LEO resources, China Satellite Network Group Co., Ltd. officially launched the StarNet Project (GW) in February 2022, planning to deploy 6,080 satellites in an extremely low orbit below 500 km (GW-A59 sub-constellation) and 6,912 satellites in a near-Earth orbit of 1,145 km (GW-2 sub-constellation), totaling 12,992 satellites. According to the filings submitted by StarNet to the International Telecommunication Union (ITU), the launch and verification of some satellites are expected to be completed by November 2027.

Although there is a clear plan for China's nascent commercial aerospace development, mass production of satellites is a significant advancement, as networking and scaling satellites are meaningful. However, apart from control technology, the development of China's reusable rockets is still constrained by liquid rocket engine technology. Currently, the recovery of liquid rockets heavily relies on engine technology, and achieving vertical recovery requires the engine to have the capability of multiple ignitions and a wide range of continuous variable thrust.

Advancing aerospace technology is a long-term journey; breakthroughs are necessary for significant progress. China should accelerate the development of liquid rocket engines to meet the needs of mass satellite launches and master the technology for recovering defunct satellites. Striving for more near-Earth satellite launches will enable China to gain more initiative in space exploration.

1.4.3 Satellite Technology Enters the Civilian Era

In recent years, the aerospace industry has drastically changed its mode of operation. New participants from the social sphere are leveraging new markets through ambitious projects to capitalize on vast business opportunities. Commercial space satellite launches have accelerated rapidly, with costs declining. Advances in satellite design, manufacturing, standardization, and shorter market entry times have significantly impacted the commercial space satellite industry, enabling faster, more flexible deployments. SpaceX is a leader in the commercial space satellite field, reusing its F9 rocket boosters while launching up to 60 lighter, cheaper satellites into LEO in a single launch. Carpooling and booster reuse have reduced launch costs, bringing the price per satellite to one million dollars.

In China, since the release of the National Medium and Long-Term Development Plan for Civil Space Infrastructure (2015–2025) in 2015, the model for developing satellite Internet in China has gradually shifted from a state-dominated model to a joint government and private enterprise approach to deploying LEO satellite Internet plans. Many private commercial aerospace companies like Landspace and Onespace have joined the communication satellite industry.

In 2018, the state-owned enterprise Guodian Gaoke deployed China's first LEO satellite IoT constellation, Tianqi Constellation. By July 2021, the first phase of the Tianqi Constellation was completed, and the second phase began in February 2022. The two major groups, China Aerospace Science and Industry Corporation (CASIC) and CASC, also initiated their own LEO communication projects, "Hongyun" and "Hongyan," at the end of 2018.

The Hongyun Project, announced in 2016 as one of CASIC's five major commercial aerospace projects, successfully launched its technology verification satellite into the intended orbit on December 22, 2018. The constellation comprises 156 LEO satellites. On December 29, 2018, the CASC launched the first experimental satellite of the "Hongyan" constellation into orbit. This constellation, consisting of more than 300 LEO satellites, plans to deploy 54 satellites by 2023 to complete the first phase and complete the overall deployment after 2024. In 2022, two Chinese satellite Internet startups, GalaxySpace and Geespace, a subsidiary of the automotive giant Geely, joined the space race to compete with the US Starlink. The former raised 1.58 billion US dollars in investment. It launched six LEO satellites in March to establish a preliminary experimental network in space, while the latter plans to build a constellation of 240 satellites. On October 19, 2022, Landspace completed the first full-system test of China's most powerful vacuum liquid methane engine, "Tianque."

Initially, national-level investment in satellite and rocket research was crucial, as individuals or private enterprises could not develop aerospace technology in terms of risk tolerance and investment in talent and research. For example, the cost of a single moon landing is equivalent to the expense of two aircraft carriers. However, with the advancement of productivity, increasing and deepening space exploration is inevitable, and the involvement of private enterprises is a natural trend. It is foreseeable that in the future, the satellite industry will develop a wide range of applications, such as LEO satellite Internet, aerospace Internet, and vehicle networking, becoming a high-tech, high-investment, high-output strategic emerging industry.

Of course, the vast market potential and development prospects also mean intense competition. On the one hand, the deployment of the satellite industry is prioritized by countries and regions like China, the US, Russia, Japan, and Europe. On the other hand, the intensifying struggle for satellite launch "permits" reflects the increasing scarcity of space orbital positions and resources. The satellite era is rapidly approaching as we transition from the ground to space.

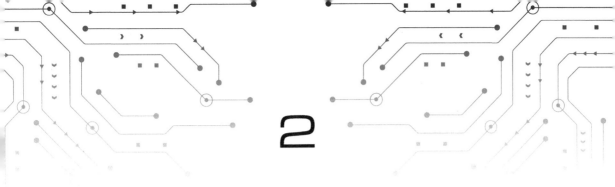

2

THE BRIGHTEST "STAR"
IN THE NIGHT SKY

2.1   The Rise of Communication Satellites

Since the Soviet Union launched the first satellite in 1957, humanity has sent thousands of satellites into space. The range of satellite applications is vast and touches everyone's life, from long-distance communication and travel navigation to watching global news and weather forecasts. For instance, many satellites today serve as navigation aids on land, in the air, and at sea. Others are used for remote sensing, such as uncovering hidden mineral resources, as well as for astronomical research and atmospheric exploration. Based on their varied applications, satellites can be categorized into communication satellites, navigation satellites, weather forecasting satellites, Earth observation satellites, and military satellites.

Currently, communication satellites are perhaps the most familiar to people. In fact, in satellite applications, communication satellites stand out significantly. Whether for television broadcasting, international phone calls, or data communication services, communication satellites are crucial as relay stations in point-to-point, point-to-multipoint, and interactive multi-point services. The majority of satellites launched into space today are communication satellites.

2.1.1 Using Satellites for Communication

The concept of using satellites for communication dates back to 1945 when Arthur Charles Clarke, a renowned British science fiction writer of the 20th century, published a famous paper in *Wireless World* magazine titled "Extraterrestrial Relays." This paper proposed the scientific concept of using communication satellites for global communication, suggesting that just three synchronous orbit satellites could achieve global communication coverage, thus giving birth to the concept of satellite communication.

In 1954, the US Navy successfully conducted transoceanic relay communication using the moon and passive balloon satellites, proving the practical value of communication satellites and satellite communication.

In 1957, the Soviet Union successfully launched the first artificial communication satellite, "Sputnik," and completed its orbit around Earth, marking the first time Earth received radio waves from a man-made satellite. "Sputnik" opened the era of active communication using artificial satellites. This landmark event by the Soviet Union, known as the "Sputnik moment," shocked the US.

In December 1958, NASA launched the world's first experimental communication satellite, "Score," into an elliptical orbit. It used onboard recording tape to achieve asynchronous telephone and telegraph communication. However, satellite communication truly became a reality with the launch of "Telstar-1" in 1962. In July of that year, AT&T launched the "Telstar-1" low-orbit satellite, establishing the first intercontinental link between the US and Europe for telephone and television services.

In December 1962, AT&T launched another low-orbit satellite, "Relay-1," which, on November 23, 1963, first achieved trans-Pacific television broadcasting between Japan and the US. The news of President J.F. Kennedy's assassination, transmitted via satellite to Japan and broadcast on television news, left a profound impression with its long-distance real-time transmission capabilities.

The world's first synchronous communication satellite was NASA's "Syncom-2," launched in July 1963. With a 30° inclination to the equatorial plane and a figure-eight movement relative to the ground, it could not yet be called a stationary satellite. "Syncom-3," launched in August 1964, positioned over the equator near the International Date Line in the Pacific, became the world's first geostationary satellite. In October of the same year, "Syncom-3" transmitted live coverage of the Tokyo Olympic Games. Thus, satellite communication was still in the experimental stage at this point.

A significant milestone in practical satellite communication was achieved on April 6, 1965, when the International Telecommunication Satellite Consortium (Intelsat) launched "Early Bird," the first semi-experimental, semi-practical geostationary satellite used for commercial satellite communication between Europe and America. This marked the entry of satellite communication into the practical phase.

A pivotal moment for satellite communication in China occurred in 1972 when China utilized leased ground stations from the US to establish satellite communication between the two countries, marking China's first use of satellite communication. Subsequently, China spent decades researching and exploring satellite communication. In 1976, China built its first fully digitalized comprehensive satellite communication Earth station. In 1984, China successfully launched its first geostationary experimental communication satellite, "Dong Fang Hong-2," inaugurating the era of using Chinese communication satellites for satellite communication.

Over the past 60 years since the launch of Telstar-1, communication satellite technology has advanced dramatically. From analog satellite communication to satellite mobile communication, from narrowband to broadband satellite constellations, we are now racing toward the future of satellite communication.

2.1.2 The Evolution and Upgradation of Communication Satellites

The advancement of communication satellites to their current level of technology was a journey marked by several challenges and generational changes. Initially, satellite communication was primarily used for TV and radio broadcasting, along with some telephone, telegraph, and fax needs. In the 1940s, British science fiction writer Arthur Charles Clarke proposed the scientific concept of global radio communication using synchronous satellites. In the 1950s, the first artificial Earth satellite was launched. By the mid-1960s, a consortium of Western countries formed the International Satellite Communication Organization, deploying the first generation of Intelsat satellites into GEO for commercial and international communication services between Europe and America. In the early 1970s, Canada was the first to launch a domestic communication satellite, "ANIK," initiating national satellite communication services and reaping significant economies of scale.

Initially, satellite communication technology primarily employed analog communication techniques, such as SCPC (Single Channel Per Carrier) and

FDM/FM (Frequency Division Multiplexing / Frequency Modulation), lacking digital circuit multiplexing and digital interfaces suited for program-controlled telephony and Internet data communication. Time Division Multiple Access required strict network clock synchronization, and the transponder bandwidth usage needed to be sufficiently flexible. With the explosive growth of program-controlled telephone services in the 1980s and the development of the Internet, experts realized that the analog technology used in satellite communication at that time could not keep up with the digital communication technology upgrades of terrestrial public communication networks. This led to the shift toward IDR (Intermediate Data Rate) technology.

IDR, a digital format upgrade, emerged in the mid-1980s as a new type of data communication service offered by Intelsat. Compared to traditional FDM/FM, IDR belongs to the TDM/FDMA (Time Division Multiplexing / Frequency Division Multiple Access) system, providing four information rates: 1.544, 2.048, 6.312, and 8.448 Mbps. Combined with Digital Circuit Multiplication Equipment, IDR could enhance satellite line transmission capacity by up to five times. For instance, an 8.448 Mbps circuit could transmit 600 voice channels.

In May 1993, China's Ministry of Posts and Telecommunications introduced IDR equipment from Canada's Spar Company, upgrading and building new satellite ground stations domestically. While IDR upgraded to digital signals and increased bandwidth, it was still insufficient to meet the rapidly growing user demands. Consequently, a new technology, VSAT (Very Small Aperture Terminal), rose rapidly in popularity. A complete VSAT system comprises a satellite transponder, a large-diameter ground main station (hub), and numerous small-diameter stations.

VSAT addressed the democratization of satellite communication, primarily through the miniaturization of ground stations. The "very small aperture" refers to diameters ranging from 0.3 to 2.4 m, small enough for one person to carry or transport by vehicle. VSAT's advantages included flexibility, low cost, diverse applications, and simple installation and operation, facilitating the widespread adoption of satellite communication.

In 1984, Intelsat first offered VSAT services. China quickly followed suit; in 1988, the China Satellite Communications Corporation introduced foreign communication equipment to build China's first VSAT communication network. The network, comprising one main station and 35 end stations, provided communication services for various governmental departments and agencies.

The development of VSAT technology laid the groundwork for numerous professional satellite communication networks, opening new vistas for satellite communication applications. However, at that time, the international Internet still needed to be fully established, so the concept of satellite Internet was still developing.

In the late 1980s and early 1990s, Motorola in the US launched the Iridium Project with an ambitious plan to "manufacture and launch 77 satellites to ensure uninterrupted coverage over every corner of the Earth's surface." The number 77 was inspired by the number of electrons in the metal element iridium. This initiative was kicked off in 1987, and four years later, Motorola established Iridium LLC specifically to advance the Iridium Project. Over a decade, at a cost of over $5 billion, Iridium LLC manufactured a total of 66 satellites. Although this fell short of the original plan by 11 satellites, Iridium LLC deemed it sufficient to begin launches. By May 1998, all 66 LEO satellites were successfully launched.

Despite the Iridium system being a comprehensive and advanced solution at the time, it had its limitations due to the technological constraints of the era. For example, the data transmission rate was only 2.4 kb/s, and the call drop rate was as high as 15%. In comparison, the first-generation terrestrial communication network GSM (1G) had a transmission rate of 9.6 kb/s, while the second-generation GSM (2G) and its enhanced versions (2.5G) achieved maximum transmission rates of up to 115.2 kb/s with call drop rates below 1%.

Moreover, due to its high costs compared to terrestrial communication networks, the Iridium system faced challenges such as high service fees, significant regional variations, and expensive mobile terminals. Consequently, both the mobile phones and the service fees were priced steeply. Iridium's mobile phones were priced at $3,000 each (equivalent to about CNY 36,000), with a per-minute call charge of $3 (equivalent to about CNY 12.5). In 2000, setting such high prices led to predictable results. Just nine months later, Iridium LLC filed for bankruptcy, as it was insolvent, with a user base not exceeding 100,000.

The Iridium project was ahead of its time, conceived during the early stages of mobile Internet development when data demands were low, and 2G networks already met communication needs. Iridium's market positioning as a replacement for existing mobile communication systems was not viable due to high tariffs and terminal prices, and it needed more commercial competitiveness against the booming terrestrial cellular communication systems, leading to its commercial failure. Although Iridium declared bankruptcy in 2000, just two years after commencing operations, the Iridium project was a success from a technological standpoint. Many of its technological concepts remain relevant today.

The launch of the Iridium project was significant, marking the beginning of the commercialization of satellite communications. During this period, other satellite communication projects emerged, such as the Orbital Communication Project (Orbcomm), which was initiated in 1995 with 35 satellites, the Teledesic project, co-founded by Abu Dhabi Investment Company, Motorola, and Boeing, and the Skybridge system, launched by Alcatel, Loral Space & Communications, and Toshiba.

In its initial phase, the Teledesic project planned to launch 840 satellites to provide the world with bidirectional broadband telecommunications services, including computer networking, rapid broadband Internet access, interactive multimedia, and high-quality voice technology. The system, known as the "Internet-in-the-Sky," was designed to use small antennae to deliver up to 100 Mb/s uplink and 720 Mb/s downlink speeds. However, due to cost constraints, the plan was scaled back in 1997 to 288 satellites spread across 12 orbital planes, with 24 satellites per plane and an additional 36 satellites as backups.

Meanwhile, in June 1991, Loral Qualcomm Satellite Services and Qualcomm proposed the Globalstar project to the FCC, planning to launch 48 satellites. Globalstar differed from Iridium in both structural design and technology, as it was not intended to form an independent network. Its primary function was to ensure that users worldwide could access terrestrial public networks via the system anytime. With handheld terminal prices comparable to those of cellular phones, the service was better suited for remote area cellular phone users, roaming users, international travelers, and countries and governments wishing to expand their communication capabilities at low cost through public and private networks. Besides, other geostationary satellite constellations, such as Astrolink, Spaceway, Cyberstar, and Euro Sky Way, faced challenges due to technological limitations and market positioning at the time. Lacking performance and cost advantages, projects like Iridium, Globalstar, and Teledesic announced bankruptcy or restructuring, with only certain services and functionalities preserved for military needs. After the failure of the first generation of Iridium, the discourse on universal satellite communication services waned as terrestrial communication systems continued to improve. However, the 21st century saw the Internet's continued exponential growth, and the demand for ubiquitous network connectivity spurred the evolution of satellite Internet capabilities. For satellite communications to develop, they had to upgrade bandwidth capacity, similar to terrestrial cellular mobile communication networks. Against this backdrop, communication satellites began evolving toward high-throughput satellite systems, marking a new era in satellite communications.

2.1.3   High Throughput Satellites (HTS)

The rise of HTS marks a significant evolution in satellite communications, offering capabilities far beyond those of traditional satellites. These advanced satellites operate in higher frequency bands such as Ku-band (12–18 GHz) and Ka-band (27–40 GHz), providing a wealth of frequency resources and substantially increased bandwidth.

One of the key distinctions of HTS is its enhanced satellite platforms. These platforms are larger and more capable, equipped with upgraded power systems that combine electric and chemical propulsion for higher energy efficiency. This advancement is crucial to support the greater power consumption required by larger bandwidths.

Additionally, HTS features a dramatic increase in the number of transponders. This expansion is akin to adding more lanes to a highway, directly translating to increased bandwidth capacity. Moreover, advancements in antenna beam technology, particularly the adoption of spot beam technology, significantly enhance signal strength. This technology focuses the signal on smaller areas, much like concentrating light with a lens, which is beneficial for increasing data rates.

A notable example of an HTS is the Thaicom 4 (IPSTAR) satellite, launched on August 11, 2005. Developed by the US company Space Systems/Loral, Thaicom 4 is regarded as the world's first HTS. Weighing around 6.5 tons with a power output of 14 kW, it was the largest commercial communication satellite at its launch. The satellite's total bandwidth capacity reaches an impressive 45 Gbps, offering uplink speeds of 4 Mbps and downlink speeds of 2 Mbps to its users.

China's foray into the realm of HTS began with the launch of Zhongxing-16 (ChinaSat 16) on April 12, 2017. This satellite employs a Ka-band multi-beam broadband communication system, boasting a total capacity of 20 Gbps. Developing and deploying such high-capacity satellites represent a significant milestone in global satellite communications, enhancing connectivity and data transmission capabilities worldwide.

2.1.4   MEO and LEO

The shift toward MEO and LEO satellites has emerged as a pivotal strategy to enhance data transmission capabilities in satellite communications. The rationale behind this move is straightforward: lower orbits, though offering a smaller coverage area, can be compensated by deploying a larger number of satellites. LEO satellites, in particular, have become a dominant force in the satellite communication industry, with over 98% of satellites launched in 2021 operating in LEO.

LEO satellites, integral to the satellite Internet, offer distinct advantages over ground communication systems. These include robust resilience to damage, wide coverage, long communication distances, rapid and flexible deployment, broad communication bands, large transmission capacity, stable and reliable performance, and an ability to operate unaffected by terrain and regional limitations. This enables seamless coverage beyond the reach of wired phone networks and ground mobile communication networks.

Compared to other satellites (GEO high-throughput satellites and MEO satellites), LEO satellites offer several benefits, including lower launch costs, superior observation capabilities, and shorter revisit cycles. With the same payload capacity, lower orbital satellites can enhance payload capabilities, improving the cost-effectiveness ratio significantly. Their lower operational altitude allows LEO satellites to rapidly reach predetermined orbits and commence operations, providing high-resolution Earth reconnaissance. This ability to closely monitor surface and low-altitude meteorological conditions often surpasses traditional satellites. Additionally, the short revisit cycle means LEO satellites can re-observe the same target within minimal time intervals, offering clearer and more frequent observations than traditional reconnaissance satellites.

Satellite Internet, evolving from LEO satellite development, constitutes a large-scale network of satellites that cover the globe, creating a significant system capable of real-time information processing. It uses satellites as relay stations to transmit microwave signals, facilitating communication between multiple ground stations and achieving global Internet connectivity. Similar to mobile and ground optical communications, satellite communication is a vital modern communication method. It boasts low latency, low cost, wide coverage, and broadband capabilities.

A satellite Internet system comprises three segments: space, ground, and user. The space segment, consisting of satellites, is responsible for receiving and relaying signals from ground stations and facilitating communication between ground stations and satellites. The ground segment includes ground stations and control stations, issuing commands to the satellites. The user segment refers to various receiving terminals primarily for sending and receiving signals. When a sufficient number of communication satellites are deployed in space, they interlace to form a satellite Internet that blankets the entire Earth, providing broadband Internet access services to ground and aerial terminals.

2.1.5 Journey to Satellite Internet

The development of satellite Internet can be traced back to the 1980s, undergoing three stages of iteration and upgrades. During the 1980s to 2000, the phase where satellite communication attempted to replace terrestrial networks, satellite Internet mainly provided services like voice, low-speed data, and IoT. However, it failed in competition with terrestrial networks due to incorrect market positioning, high technical complexity, excessive investment, prolonged development cycles, and weak system capabilities.

Entering the 21st century, satellite communication gradually supplemented terrestrial networks. In this stage, Internet satellites primarily complemented and extended terrestrial communication systems while providing mobile communication services to aviation, maritime, and other users under extreme conditions.

Today, the development of satellite Internet has entered its third phase: integrating satellite and terrestrial communication networks. In this era, companies like OneWeb and SpaceX, leveraging innovations in rocketry, material processes, and millimeter-wave communication technologies, are leading the construction of new satellite Internet constellations. Represented by Starlink and OneWeb projects, satellite communication is no longer just a supplement to terrestrial networks. Still, it is becoming an integral part of them, merging and evolving alongside terrestrial networks.

Currently, in terms of latency, satellite Internet has achieved delays of several tens of milliseconds, comparable to 4G networks but still significantly behind the sub-10 ms delays of 5G networks. This indicates that satellite Internet cannot yet meet the low latency requirements of applications like autonomous driving and telemedicine.

This limitation arises from the scarcity of spectrum resources. Enhancing system capacity necessitates increasing the number of satellites and employing multi-beam technology per satellite, using frequency reuse principles to boost system capacity. In bandwidth terms, although the peak communication rate of individual satellites has reached 20 Gbps, comparable to 5G base stations, the sheer number of terrestrial base stations outnumbers networking satellites, thus limiting satellite Internet in accommodating a massive number of terminals. In other words, satellite Internet does not currently compete with terrestrial networks like 5G.

Looking ahead, satellite Internet remains a crucial infrastructure for building an integrated terrestrial and space information network. Although it shares advantages like low latency, low cost, broad coverage, and wide bandwidth with terrestrial fiber optics and 5G, satellite Internet's greatest advantage lies in its independence from geographical constraints and lower reliance on ground infrastructure. In the future, with its broad coverage and low cost, satellite Internet can not only effectively complement fiber Internet and mobile Internet but also aspires to build a truly integrated space and terrestrial information network, achieving seamless global coverage.

2.2 Satellite Complements 5G

While the construction of 5G terrestrial communication networks continues vigorously, research into the 6G network, characterized by the integration of space and terrestrial systems, has also commenced. Crucially, the construction of satellite communication networks is vital to transitioning from today's 5G era to the integrated 6G future.

2.2.1 Addressing the "Last Mile" on the Ground

On August 29, 2023, Huawei's Mate 60 Pro phone, featuring satellite call capability, garnered significant attention. According to Huawei, the Mate 60 Pro is the world's first mass-market smartphone with satellite call functionality, ensuring connectivity even in areas without terrestrial network signals.

Following Huawei, Apple's iPhone 14 and 15 series also integrated satellite communication features. Additionally, Samsung is collaborating with Iridium Communications to provide satellite communication for its upcoming Galaxy series. By the end of 2023, other Chinese smartphone manufacturers are expected to support satellite call features. Satellite communication is rapidly becoming a standard feature in high-end smartphones and is extending to other consumer electronics like smartwatches.

Satellite calls have recently been a hot concept in the smartphone industry. This enthusiasm stems from the potential of satellite Internet. Most communications rely on terrestrial systems, transmitting information via ground stations or infrastructure through mediums like cables or radio waves. Satellite communication, however, transmits information to satellites in Earth's orbit and then to the desired location.

The primary advantage of a satellite Internet constituted by numerous communication satellites is its extensive signal coverage, addressing the perpetual issue of coverage gaps in traditional communication.

Despite the rapid development of China's 5G technology and its widespread coverage in urban and rural areas, terrestrial mobile communication networks cover less than 40% of the country's land area.

Globally, Internet backbone networks have received substantial investment and attention, preparing them for various broadband services. However, the "last mile"—the connection from the backbone to the user—has lagged due to various factors, becoming the Internet's bottleneck. Only by resolving the "last mile" issue can the Internet achieve its ultimate goal of being high-speed, interactive, and personalized.

Currently, global Internet data is primarily transmitted through underground or underwater cables. The high cost and limited coverage of cable deployment are major obstacles to Internet expansion. To truly achieve a connected world, satellite Internet, capable of global coverage, is essential.

2.2.2 The Anticipated 6G Era

With its vast coverage and long communication distances, satellite communication shows immense advantages in large-area, sparse routing, and seamless mobile communications. Satellite Internet, constituted by numerous communication satellites, is crucial for future communications.

Satellite Internet will complement terrestrial networks. Integrating satellite communication networks with terrestrial ones to form a unified information network can achieve global coverage, filling the gaps in existing Internet networks and meeting the connectivity needs of remote areas and users by air and sea. This is vital in various aspects, including emergency response and scientific exploration.

In emergency response, satellite communication can be lifesaving, maintaining connectivity even in remote areas with no signal, like deserts or dense forests. For instance, a hiker once got lost in heavy snow and used a Huawei Mate50 phone to send a distress signal via satellite, leading to a successful rescue. In another case, a traveler with a broken-down car in a low-signal area used a Huawei P60 Pro to send their location via satellite, aiding in timely assistance.

For ocean operations and scientific expeditions, satellite Internet is the only solution for communication needs, as traditional ground stations cannot be deployed at sea or in the air. Satellite Internet can connect ships at sea with data centers ashore, optimizing energy efficiency and operational monitoring.

Additionally, satellite communication can meet the IoT requirements by enabling high-speed information access from the Internet and facilitating environmental and object monitoring and data collection. It provides a more convenient and economical solution for remote area connectivity, enabling applications like power grid monitoring, geological surveys, forestry monitoring, UAV data transmission, maritime buoy data collection, offshore container tracking, crop monitoring, and surveillance of rare animals in uninhabited areas.

Satellite Internet extends and complements terrestrial Internet and strengthens it, accelerating the progress toward multimedia communication and personalized services. As a new network system and information service, satellite Internet crosses over telecommunications, the Internet, and broadcasting industries, playing a crucial role in the convergence of these sectors.

As satellite Internet approaches, the outline of the 6G era is becoming more apparent. Although 6G currently lacks a precise definition, its communication standards are expected to exceed those of 5G with higher rates, broader bandwidth, and lower latency. Building on 5G, 6G aims to support efficient connectivity among intelligent entities, achieving seamless global coverage by integrating space, air, and terrestrial networks. Satellite Internet is poised to become a vital solution for global satellite communication network coverage in the 5G and 6G eras and a key trend and strategic high ground in the concurrent development of the aerospace, communication, and Internet industries.

2.3 From Terrestrial Interconnectivity to Collaborative Space-Earth Networks

Despite the rapid development of 5G and the forthcoming arrival of 6G, the essence of 5G lies in its capacity to interconnect everything at higher speeds and quality, propelling a multitude of industries into a new era. Parallel to the 5G boom, satellite Internet is making significant strides toward becoming a tangible aspect of human life. The era of large-scale space exploration promises to be even more magnificent than the Age of Exploration, forging unprecedented connections and potentially shaping new worldviews and cosmic perspectives.

2.3.1 The Prime Time for Communication Satellites

Satellite communication, utilizing orbiting satellites as relay stations to reflect or transmit radio signals, has evolved significantly from military origins to civilian use and high to LEO. Initially used by the US military for naval communication, satellites were later leased to civilian organizations for maritime emergency services. In the 1990s, civilian organizations began launching their geosynchronous satellites, marking the commercialization of satellites. However, due to their distance from Earth and weaker signal transmission, geosynchronous satellites presented issues with call quality and required more sophisticated and costly terminals.

Companies proposed constellation plans for medium and LEO satellites to overcome these challenges, exemplified by the "Iridium" and "Globalstar" projects. These plans involved networks of dozens of satellites covering the globe. In the 21st century, with the maturity of mobile Internet, satellite Internet gradually entered public awareness. Giant constellation plans by major companies aimed to cover global Internet support, escalating satellite numbers from dozens to thousands and reducing the orbital altitude to 300 km, as seen in the "OneWeb" and "Starlink" projects.

During this evolution, satellite launch costs have steadily decreased, and manufacturing technology is maturing. Since 2000, new technologies and increased competition have significantly reduced the costs of satellite launching and manufacturing. From 1970 to 2000, the average satellite launch cost was about $18,500/kg. The emergence of new launch providers like SpaceX has reduced this cost to approximately $2,720/kg, an 85% reduction. Simultaneously, satellite weights have also decreased. For instance, the first Iridium satellites launched in the late 1990s weighed 689 kg each, whereas SpaceX's Starlink satellites now weigh around 250 kg.

Regarding manufacturing technology, satellites, and their production processes have become more advanced. Companies increasingly adopt modular satellite designs, utilizing standardized bus technologies and smaller, more advanced components. Many use electric propulsion systems to reduce weight and costs, gaining a competitive edge. Specialized factories for building these massive constellations have emerged, reducing costs and accelerating the pace of mass production. For example, the joint venture between Airbus and OneWeb, OneWeb Satellites, can manufacture two daily satellites.

As we transition from 5G to the era of collaborative space-Earth networks, the role of satellite communication becomes increasingly pivotal, not just as a complement to terrestrial networks but as a crucial component of the impending 6G era. This convergence of terrestrial and space networks is set to redefine connectivity and communication, opening new frontiers in global interconnectivity.

2.3.2 The Dawn of the Satellite Age of Exploration

At the turn of the last century, LEO satellite communication systems, such as the Iridium and Globalstar voice communication systems, existed. Several mid- and low-orbit communication systems plans, like Microsoft's Teledesic constellation and Europe's "Skybridge" constellation concept. However, the high-income market envisioned by Iridium did not materialize, and its high-pricing strategy led to its failure. As a result, most subsequent constellation concepts remained conceptual and were not implemented.

Two decades later, with in-depth research in satellite communication architecture, frequency allocation, access methods, interference, signal attenuation, pricing, and gradually cultivated application demands, a new generation of LEO broadband satellite systems has begun construction. The current LEO broadband constellations have clearly defined their market positioning toward the general consumer group. The lower launch prices, brought about by low-cost small satellite technology and new rocket developments, support this market positioning, making satellite broadband services affordable for average consumers.

Globally, significant LEO constellation constructions (consisting of more than 500 orbiting satellites) include Starlink, OneWeb, Boeing Constellation, and Samsung Constellation, with other LEO communication constellation plans not exceeding 300 satellites.

SpaceX's StarLink project plans to launch 42,000 high-, middle-, and low-orbit satellites to achieve global networking in the current LEO satellite communication system. The first phase involves launching 4,425 satellites into LEO ranging from 1,110 km to 1,325 km, covering global wireless network signals. The remaining 7,518 satellites will be sent to an even lower orbit of 335 km to 346 km, primarily providing sufficient bandwidth and low-latency data transmission for densely populated areas.

Additionally, the "Skybridge" constellation, initially proposed by France's Alcatel in the early 21st century, was stalled due to the bankruptcy of Iridium. However, the system design and frequency application work had been completed. OneWeb later acquired these assets to restart the "Skybridge" construction, which mainly consisted of a large-scale constellation using the Ku band to be compatible with existing satellite communication terminals. OneWeb's satellites have already occupied some orbital resources and have official authorization from the ITU for part of the Ku band resources. OneWeb's market mainly focuses on constellation Internet, covering areas globally that currently lack Internet services.

In China, the Hongyan Constellation by the CASC is a major national commercial aerospace project operated by China Spacesat Co., Ltd., a joint venture with China Telecom, CEC, China Reform Holdings Corporation Ltd., and other companies. The first phase of the "Hongyan Constellation" plans an investment of CNY 20 billion to establish a communication network of 60 satellites by 2022. The second phase aims to deploy 320 satellites by 2025, building an integrated satellite mobile communication and space Internet access system covering "sea, land, air, and space," enabling global Internet access and providing various mobile communication services.

By the end of 2018, the "Hongyan Constellation's" first experimental satellite, "Chongqing," was successfully launched. It conducted in-orbit verification of key technologies for space-based Internet systems and demonstrated capabilities in mobile communications, broadband Internet, the IoT, and navigation enhancement while actively exploring business models.

Simultaneously, CASIC commenced the implementation of the Hongyun Project, planning to deploy 156 satellites at an altitude of 1,000 km by 2022 to construct a global mobile broadband network capable of seamless worldwide coverage. The Hongyun Project was divided into three phases: "1 + 4 + 156." The first phase aimed to launch the first technology verification satellite before 2018 to achieve key technological verification on a single satellite basis. The second phase, by the end of the 13th Five-Year Plan period, aimed to launch four service test satellites to form a mini constellation, allowing users to have an initial experience of the service. The third phase, by the end of the 14th Five-Year Plan period, aimed to have all 156 satellites operational in the constellation, completing the construction of the service constellation.

The primary user base for the Hongyun Constellation includes clustered user groups such as aircraft, ships, trucks, remote field areas, working teams, and remote villages and islands. It also serves the drone and autonomous driving industries. With its extremely low communication latency, high-frequency reuse rate, and truly global coverage, the Hongyun Constellation meets the broadband Internet access needs of both China and international Internet underdeveloped areas and the needs of large-scale user units sharing broadband access to the Internet. It also meets the demands for applications requiring high real-time interaction, such as emergency communications, sensor data collection, and the remote control of unmanned equipment and industrial IoT.

Beyond the LEO satellites developed by the defense and industrial systems, private companies within the domestic aerospace market, such as Galaxy Space, are also actively positioning themselves in the low-orbit satellite communication market. Galaxy Space is committed to agile development and rapid iteration models, scaling the production of low-cost, high-performance small satellites to create a leading global LEO broadband communication satellite constellation, establishing an integrated global terrestrial and satellite communication network.

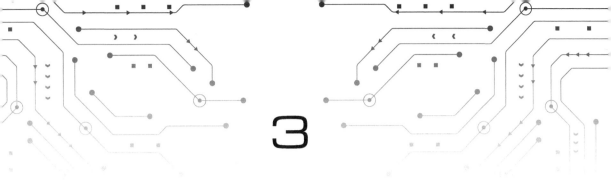

$$3$$

SATELLITE COMMUNICATION—
THE BATTLE FOR TERRITORY

3.1   The Scramble for Satellite Communication Entry Tickets

A new space race, dominated by information communication, has begun. Communication has become a core competitive force in business, including military intelligence. Particularly under the rapid development of satellite Internet, building new satellite communication infrastructures is a trend, with LEO satellite communication becoming a fiercely contested international arena. Notably, US enterprises like SpaceX are accelerating the deployment of LEO satellites, intensifying the space race. The urgency stems from the finite and non-renewable nature of space orbits and spectrum resources. The international principle of "first come, first served" and the growing importance of the LEO satellite communication industry make it a focal point of competition among major powers, both for possessing orbital resources and future communication control.

Globally, many countries are focusing on LEO satellite communication. A wave of LEO satellite communication development, driven by technological innovation and commercial capital, is sweeping across the globe. Seizing the rhythm in this wave presents a significant challenge for the satellite industry's development in each country. It is a crucial opportunity to advance into the integrated "space, sea, land, and air" 6G era.

3.1.1 Non-renewable Spectrum Resources

Establishing any satellite system requires an appropriate radio frequency and an orbital position, essential for the system's function. In other words, a satellite system can operate with frequency and orbital support. For communication satellites, this is equally true.

Satellite frequency mainly refers to the radio frequency range used for transmitting and receiving information in satellite communication systems. This frequency range determines the transmission characteristics of satellite signals, directly affecting the performance and effective operation of the satellite communication system. The working frequency range for satellite communication is 1 GHz to 40 GHz (1 GHz = 1,000 MHz) in the microwave band of the electromagnetic spectrum. Based on different frequencies, they can be divided into L, S, C, X, Ku, Ka, and Q/V bands, with each band serving different purposes. With significant atmospheric absorption loss, the K band is unsuitable for satellite communication. Hence, L, S, C, X, Ku, and Ka are the commonly used bands (see the table below for details).

Bands	The working frequency range/GHz	Wavelength/cm
L	1–2	30–15
S	2–4	15–7.5
C	4–8	7.5–3.75
X	8–12	3.75–2.5
Ku	12–18	2.5–1.67
K	18–27	1.67–1.11
Ka	27–40	1.11–0.75

L and S bands are used for satellite mobile communication, like the handheld satellite phones often seen on TV. With its lower frequency, the C band offers broad signal coverage and minimal weather impact and is commonly used for corporate network communication, international lines, TV content distribution, etc. The X band is regulated, typically for government and military use, not for civilian commercial purposes. With a relatively higher frequency, the Ku band is prone to signal fluctuation due to antenna influence. However, its stronger signal compared to the C band allows for much smaller ground-receiving antennae, suitable for direct-to-home satellite TV broadcasting. The Ka band, recently gaining popularity, has characteristics similar to the Ku band. Despite being more susceptible to weather, its larger bandwidth, higher signal strength, and smaller antennae make it valuable for high-speed satellite communication, gigabit broadband digital transmission, HDTV, satellite news gathering, VSAT services, direct-to-home services, and personal satellite communication. Importantly, satellite spectrum resources are limited. In the early 21st century, a struggle for satellite frequencies once played out, particularly for navigation satellites. Europe and China proposed the Galileo and BeiDou satellite navigation systems, respectively, leading to a classic battle over frequencies.

According to the ITU, there are only five frequency segments allocated for navigation: 1,164–1,215 MHz, 1,215–1,240 MHz, 1,240–1,260 MHz, 1,260–1,300 MHz, and 1,559–1,610 MHz. The US GPS and Russia's GLONASS had already occupied central frequencies around 1,176 MHz, 1,227 MHz, 1,246 MHz, and 1,602 MHz. This left only 1,260–1,300 MHz and parts of the other four segments for China and Europe.

Both BeiDou and Galileo's high-precision services center around 1,207.14 MHz. China applied for this frequency on April 17, 2000, and Europe on June 5, 2000. According to regulations, they needed to utilize this frequency by April 17, 2007, and June 5, 2007. Eventually, a compromise was reached in 2015, and both systems now share this frequency.

Today, a similar scenario is unfolding, not for navigation satellite frequencies but for communication satellite frequencies.

The currently usable L, S, and C band resources for global coverage are nearly exhausted. The widely used Ku and Ka bands for GEOs broadband satellites face saturation, with Ku band nearing full capacity and Ka band extensively utilized. As constellations require frequency gaps to prevent interference, coordination.

3.1.2 The Rush for Orbital Space: A Battle for Non-renewable Resources

Beyond spectrum resources, near-Earth orbital space has become a highly coveted asset, evolving into a non-renewable resource in the aerospace arena.

Although it seems theoretically possible to launch countless satellites between the Earth's surface and the GSO, making near-Earth orbit appear boundless, especially against humanity's current spacefaring capabilities, space is filled with risks. Every spacecraft in LEO travels at speeds exceeding 7 km/s, carrying tremendous kinetic energy and destructive potential. Exceeding a threshold of uncontrollable satellites in space could trigger a chain reaction, leading to massive satellite collisions and an abundance of space debris, devastating the entire LEO satellite fleet. Hence, limiting the number of space satellites is necessary for their safe operation.

According to a report by China's CCID Consulting, Earth's near-Earth orbit can accommodate approximately 60,000 satellites. Predictions suggest that by 2029, about 57,000 LEO satellites will be deployed. Currently, SpaceX's Starlink alone plans for 42,000 satellites.

The ITU allocates orbital and frequency resources using two methods: the "planning method," which divides resources equally among countries (favoring developing nations), and the "coordination method." The latter involves three steps: initial declaration of satellite network information to ITU, negotiation and technical adjustment with concerned parties if interference issues arise, and final allocation and registration of orbit and frequency by ITU if requirements are met. Once registered, the declared orbital path enjoys internationally recognized rights, but they expire if the satellite is not operational within seven years, necessitating a re-run of the coordination process.

Currently, the main international practice for distributing satellite frequency and orbital resources follows a "first come, first served" approach. Once a nation occupies an orbital space, it gains "permanent use rights," making the resources unavailable to others.

Before 2019, ITU regulations were lenient, considering a satellite network operational even if only one of its satellites was in use. This led to the prolonged occupation of certain satellite frequency and orbital resources. To prevent "hoarding frequencies," ITU imposed a new rule: networks expiring after January 1, 2021, must deploy 10% of their satellites within two years, 50% within five years, and 100% within seven years, or scale down their constellation. Afterward, the number of satellites in orbit must remain at 95%; if this is not met for six consecutive months, a replenishment plan must be reported to ITU.

Under these rules, several Asia-Pacific countries, including Japan, India, South Korea, and Malaysia, independently or collaboratively manufacture communication satellites to secure orbital resources. Satellite "collisions" among nations are not uncommon. The race for satellite frequency and orbital resources, a struggle for space superiority, has become a hot topic in the world of satellite development.

The vastness of space also harbors ancient human dreams. The era of satellite Internet, akin to a grand space age, promises to be even more monumental than the great age of maritime exploration. Currently, the struggle in satellite communication is also a battle for orbital resources. Humanity's future inevitably shifts from terrestrial interconnectivity to a combined terrestrial and celestial collaboration. The fight for entry tickets has just begun in this grand space endeavor.

3.2 The Global "Starlink" Contest among Major Powers

Given the finite nature of satellite spectrum and orbital resources, the struggle for these assets is inevitably fierce. Currently, nations have fully recognized the strategic importance of frequency and orbital resources and are vigorously vying for them. Especially since the feasibility of the LEO satellite Internet business model was validated by Starlink, countries worldwide have accelerated the launch of their own "Starlink projects."

3.2.1 The US: Leading the Global "Starlink" Race

The US boasts advanced aerospace technology and facilities. Since the 1960s, the US has embarked on large-scale space programs, successfully launching numerous satellites. Subsequently, the US continued to increase investments in space, building a series of state-of-the-art launch bases and R&D facilities, forming a complete aerospace industry chain. This created favorable conditions for the US to launch even more satellites.

In satellite communication, the US has continuously introduced top-level strategic policies. Since the 1950s, each US administration has introduced new national space policies. Under the Trump administration, the National Space Council was reestablished, and the National Space Strategy was formulated for the first time. Four presidential space policy directives were signed, further strengthening guidance and support for the aerospace industry. Regarding regulatory systems, the US has the earliest, most mature, and most comprehensive satellite communication regulations globally.

Since the National Aeronautics and Space Act of 1958, the US has enacted the Communications Satellite Act, the Orbital Act, and other single-issue laws to regulate and encourage the development of the commercial satellite communication industry. In the launch domain, laws like the Commercial Space Launch Act, Commercial Space Law, Launch Services Purchase Act, Commercial Space Launch Competitiveness Act, and Encouraging Private Aerospace Competitiveness and Entrepreneurship Act have effectively regulated and promoted private enterprise participation in satellite launching. In March 2022, the US released a consultation document, Enhancing the Cybersecurity of Satellite Communications Network Providers and Customers, proposing specific measures to strengthen satellite communication network security.

In March 2023, the US National Science and Technology Council published the National Near-Earth Orbit Research and Development Strategy, outlining America's vision to lead in near-Earth orbit future R&D. This includes driving scientific and technological progress, strengthening interdepartmental and public-private cooperation, promoting market sustainability, expanding international partnerships, engaging the public, and supporting the sustained development of the US space workforce. It set policy goals and priorities for the US in the LEO satellite domain. In November of the same year, the National Spectrum Strategy was released to modernize spectrum policy and efficiently use spectrum resources by establishing four strategic directions and twelve specific strategic objectives. A key content was establishing and implementing a national framework for spectrum policy cooperation, solving the difficulty in acquiring spectrum resources through dynamic spectrum sharing.

The US satellite communication industry's development is characterized by flourishing commercial applications, with the Starlink project being the most iconic. SpaceX proposed the Starlink project in 2015 and launched the first batch of 60 satellites with the Falcon 9 rocket on May 23, 2015. By the end of December 2023, SpaceX had launched over 5,000 "Starlink" satellites, nearly two-thirds of the LEO satellite orbit resources, securing an absolute industry-leading position.

In addition to SpaceX's Starlink, Amazon's Kuiper project aims to invest billions of dollars to launch 3,236 near-LEO satellites, later applying to increase to 4,538 satellites. Boeing plans to send 147 satellites into near-Earth orbit within nine years, later increasing to 5,789 satellites. Rocket manufacturer Astra is also actively deploying a space network of 13,600 satellites. Moreover, several US startups have submitted applications to deploy hundreds to thousands of satellites.

Besides commercial applications, the US military satellites are diverse and technologically advanced. The US military communication satellites are mainly categorized into broadband, narrowband, and protected types. The broadband satellites include three generations of the Defense Satellite Communications System (DSCS). The current Wideband Global SATCOM (WGS) satellites are replacing the DSCS, with ten WGS satellites launched and the construction essentially complete. The narrowband communication system has developed the "Fleet Satellite," "Leased Satellite," and "Ultra High-Frequency Follow-On" satellite series, with the MUOS system satellites replacing the latter since 2012. In the protected satellite domain, the US has developed the "MILSTAR" series, replaced by "Advanced Extremely High Frequency" satellites since 2010, with four successfully launched.

Another characteristic of the US satellite communication industry is the trend of dual use for military and civilian purposes. In April 2022, the US released the Space Command Commercial Integration Strategy to promote commercial space capabilities for defense empowerment. In December, SpaceX announced the "Starshield" project to use LEO satellites for Earth observation, secure communication, and payload hosting services for the Department of Defense. The US military also uses commercial satellites like "Starlink" and "OneWeb" for communication trials and networking commercial and military satellites for tactical and strategic communication.

Traditionally, US military communication heavily relied on GSO satellites, which could not afford the loss of high-value satellites. To mitigate risks, the US military proposed the concept of "mosaic warfare," replacing high-value satellites with low-value, networked satellites with self-organizing capabilities. This approach ensures that communication remains intact even if enemies destroy some nodes. Given these advantages, LEO communication satellites have begun to play a prominent role in the military domain.

LEO communication satellites, characterized by high transmission rates, wide coverage, strong survivability, small ground terminals, and easy deployment to the frontlines, have become a core communication component in just a few years. The US military even plans to embed them into military communication networks to provide communication support for US forces and allies globally. Besides communication capabilities, the "Starshield" scheme based on the second-generation "Starlink" can provide comprehensive services like remote sensing, communication, and payload hosting. The first- and second-generation "Starlink" practice indicates that LEO communication satellites will inevitably become a vital infrastructure for major military powers to strengthen intelligence, command, and strike chain integration.

With modern warfare evolving toward unmanned and intelligent directions, LEO communication satellites will also become essential logistics support facilities for future unmanned combat clusters. The entire kill chain, from detection and location to assessment, is fundamentally based on communication. Evidently, satellite communication is key to accurately and efficiently completing targets in the kill chain. The kill chain currently relies on high-orbit space-based satellites, often leading to delayed kill chain closure in wartime. With numerous unmanned combat nodes joining, the pressure on the communication system to operate normally and manage damage will further increase. LEO communication satellites can connect unmanned equipment across different regions and systems, not only speeding up the kill chain closure but also significantly reducing data interaction time for unmanned combat clusters.

Additionally, LEO communication satellites can serve as computing nodes. Existing unmanned combat clusters consolidate massive battlefield information into command-and-control systems during combat missions. This requires strong communication links and robust computing systems to discern intelligence's authenticity, timeliness, and value. Upgrading LEO communication satellites to intelligent computing nodes means that combat information obtained by unmanned combat clusters can be assessed and processed directly on the battlefield, greatly enhancing the overall efficacy of the clusters. Therefore, relying on more efficient LEO satellites to reconstruct the kill chain will become an inevitable trend. SpaceX has already clearly demonstrated this capability in the wars in Russia-Ukraine and Israel-Palestine.

It is foreseeable that in modern warfare, with disruptive technologies like LEO satellite systems applied in war, more private enterprises and civilian forces will participate in future combat operations.

3.2.2 Europe: Challenging "Starlink" with "OneWeb"

Following closely behind the US, Europe has given high priority to the development of satellite communication at the government level, seeing it as a key lever to enhance international influence and discourse. Clear strategic plans have been formulated, with continuous budget investments and active international collaboration to vigorously promote satellite communication development.

As early as 2008, the European Union formally initiated the approval process for pan-European satellite mobile communication services, streamlining procedures for satellite mobile communication services across the EU. In 2011, the European Commission explicitly proposed integrating member-state resources to develop pan-European satellite mobile communication at the EU level. The European Space Strategy issued by the European Commission in 2016 emphasized advancing European space integration. To implement the European Space Strategy, the EU 2021–2027 Long-Term Budget Proposal released in 2018 proposed the European Space Program, which calls for ensuring high-quality, up-to-date, and secure space-related data and services, providing more convenient support for commercialization.

In November 2023, EU member states reached a political agreement on the 2023–2027 EU Secure Connectivity Plan with a total budget of 2.4 billion euros. This plan, named "IRIS," is the EU's third flagship space program after Galileo and Copernicus, aiming to build a resilient, interconnected, and secure satellite infrastructure system. The plan has two parts: establishing a sovereign space-based secure connectivity system to provide secure, autonomous, reliable, and cost-effective satellite government communication services for government users and supporting the protection of critical infrastructure crucial to the economy, environment, security, and defense; and providing commercial services to the private sector, promoting further development of global high-speed broadband, seamless communication, and enhancing cohesion among member states to achieve the European Digital Decade's goals.

In addition to government support for satellite communication, Europe is accelerating infrastructure construction and commercial applications. Leveraging the telecommunications company OneWeb, the UK launched the OneWeb constellation plan, which has become the world's second-largest communication satellite constellation.

OneWeb, initially known as World Vu Satellites Ltd., was founded in 2012 by Gregory Thane Wyler, the founder of O3b. In 2014, OneWeb proposed a constellation plan of 648 small satellites. Subsequently, it acquired the necessary frequency and orbit resources from the nearly bankrupt "Skybridge" company and registered with the ITU. In 2015, OneWeb officially entered the constellation construction phase.

In 2017, OneWeb applied to the FCC to increase the number of satellites to 720 in near-Earth orbits at 1,200 km and 1,280 in MEOs at 8,500 km. In early 2020, OneWeb faced bankruptcy due to the COVID-19 pandemic but was later jointly acquired by the UK government and India's Bharti Enterprises and gradually recovered from 2021. According to OneWeb's plan, it will deploy at least half of the planned satellites by August 2026 and complete the entire OneWeb constellation by 2029.

OneWeb adopts an open architecture, enhancing system capacity by adding new satellites. The constellation construction is divided into three phases. The first phase involves launching 648 Ku/Ka band satellites distributed at an altitude of 1,200 km in 18 orbital planes, with approximately 40 satellites in each plane and an orbital capacity of 7 Tbit/s, providing users with peak broadband services of up to 500 Mbps and a ground-to-satellite delay of about 50 ms. The second phase adds 720 V-band satellites to form a "sub-constellation" at the same orbital height (1,200 km), with a constellation capacity of 120 Tbit/s. The third phase adds 1,280 V-band satellites operating in higher MEO (8,500 km), with a constellation capacity of 1,000 Tbit/s, and dynamically allocates traffic between LEO and MEO based on service demand and data flow in the coverage area.

In March 2023, OneWeb successfully launched 36 first-generation constellation satellites by the Indian space startup NewSpace India, bringing the actual networked satellite count to 618, nearly achieving global service capability. OneWeb's constellation plan has thus become the world's second-largest communication satellite constellation after Starlink.

Currently, both OneWeb and Starlink have entered the full deployment phase, but there are significant differences in their system architecture and production operations. Starlink adopts a "sky-satellite and sky-network" architecture, using satellites as network transmission nodes and establishing a high-speed broadband communication network through inter-satellite links, allowing users to directly access the satellite Internet network without going through a ground system. In contrast, the OneWeb constellation uses a "sky-satellite-ground network" architecture, with satellites serving as channels connecting user terminals and gateway stations. The constellation has no inter-satellite links, and the system provides global service capability through globally distributed ground stations.

Regarding production and operation, SpaceX handles Starlink's satellite development, production, launch, and constellation operations. SpaceX's rocket and reusable launch technology significantly reduce the launch costs of the Starlink constellation. On the other hand, OneWeb focuses on vertical integration of the entire industry chain, partnering with leading companies like OneWeb, SoftBank Group, Airbus, and Hughes Network Systems to form an integrated interest group.

Notably, OneWeb also possesses large-scale satellite manufacturing capabilities. In 2017, it began constructing the Florida factory near Kennedy Space Center, with two satellite production lines, and the Toulouse production line in France started producing the first batch of satellites. OneWeb's satellite production line has a capacity of 60 satellites per month.

Regarding commercial applications, on September 29, 2023, OneWeb announced its official merger with French satellite company Eutelsat to become Europe's largest satellite company, entering full competition with Starlink. The merger allows OneWeb to acquire sufficient funds to complete new network construction and technological updates. Additionally, the new company is expected to combine the advantages of Eutelsat's high-throughput GEO satellites with OneWeb's LEO satellites to provide customers with low-latency and comprehensive coverage of high-density networks.

On November 21, OneWeb India received authorization from the Indian National Space Promotion and Authorization Center to provide commercial satellite broadband services from the European Communication Satellite Organization OneWeb in India. OneWeb India is the first organization to receive this authorization. On November 23, Hanwha Systems announced a supply agreement with Eutelsat OneWeb for LEO satellite communication, providing ultra-high-speed, low-latency communication connections for every corner of South Korea.

Beyond the UK, Germany is pushing the LEO satellite network layout through startups like Rivada Space Networks and KLEO Connect. Rivada Space Networks announced in March 2022 its plan to start constructing a constellation of 600 LEO satellites within 18 months, with deployment beginning in 2024 and expected completion by mid-2028. These satellites will combine satellite and terrestrial capabilities to provide high-speed, low-latency global network coverage.

Meanwhile, the SES operates the world's only medium orbit communication satellite constellation, O3b. The second-generation constellation, O3b mPOWER, is currently under construction, with four satellites already launched.

In summary, Europe is actively advancing in satellite communication development through government support, infrastructure construction, and commercial applications. OneWeb, emerging from the UK and now competing directly with Starlink, exemplifies Europe's ambition in the global communication satellite arena.

3.2.3 Russia: Comprehensive Deployment of Satellite Communication

Russia has always been a significant player in the field of space exploration. The notable "Star Wars" competition with the US during the Soviet era showcased Russia's remarkable space capabilities. Inheriting most of the Soviet Union's legacy, Russia is regarded as a nation with profound expertise in space technology.

Regarding policy and regulation support, Russia's legal system concerning satellite communication mainly consists of the Federal Law on Space Activity along with numerous presidential decrees, government orders, and industry-related regulations and institutional arrangements. The Federal Law on Space Activity explicitly states that space technology and activities are "national top-priority development projects" to enhance Russia's economic, technological, and defense capabilities through space technology. Russia has issued multiple space strategy plans, clarifying the key development directions and phased deployment tasks in the space sector, significantly guiding the development of its space industry.

In recent years, Russia has primarily launched military communication satellites to complete its domestic high-performance, multi-purpose military communication satellite series. Currently, Russian military communication satellites in orbit are mixed into the "Cosmos" series, mainly divided into LEO, HEO, and GEO satellites, providing various communication and command and control services for Russian armed forces.

The LEO communication satellites include the Strela-3 and Rodnik types. "Strela-3" is a store-and-forward communication satellite, not supporting real-time communication, mainly used for transmitting military intelligence. HEO satellites include the "Molniya" series, used for both military and civilian purposes, and its successor, "Meridian," undertaking strategic communication tasks for the Russian government and military departments, focusing on military command, control, communication, and civilian communication and surveillance services. The GEO satellites include "Raduga" and "Gonets" models. "Raduga" transmits data, phone calls, television, and specialized information, providing encrypted secure channels for high-level military communication. "Gonets" belongs to data relay satellites.

Regarding civil and commercial communication satellites, Russia has developed the "Messenger," "Express," and "Yamal" satellite series. The "Messenger" system is also a dual-use satellite, providing personal communication and data exchange services, ensuring communication links between Russian government departments and remote areas. The "Express" series of high-performance communication satellites provide satellite broadcasting, broadband access, and mobile communication services for Russia and surrounding regions. Russian Satellite Communications Company has signed cooperation agreements with satellite service providers in the Middle East, such as Horizon Sat, Chronosat GmbH, and Romantis, allowing them to use satellites like Express-AM6, AM7, and AM22 to provide communication services in the Middle East, Central Asia, South Asia, and other regions, expanding the commercial market.

Additionally, Russia initiated the "Sphere" project in 2018, planning to build a satellite constellation comprising over 600 satellites in the coming years. The "Sphere" constellation is expected to operate at an altitude of 870 km, higher than the operational altitude of the US "Starlink." By operating various satellites at different altitudes, the "Sphere" constellation will have diverse functions, including satellite communication services and Earth observation, making it "the most modern space communication and monitoring system" for Russia. According to Russian government plans, government funding of 18 billion rubles is expected in 2023 and 2024, reducing to 8.5 billion rubles in 2025.

3.2.4 China: Accelerating the Creation of China's Version of "Starlink"

Given the global satellite communication industry's "land grab," which encompasses not just commercial competition but also the struggle for technological supremacy and national defense strategy, China has hastened its pace in the satellite communication sector, driven by the scarcity of resources and the proactive moves of the US.

First, China has integrated the development of satellite communication into its national strategic planning. In April 2020, China's National Development and Reform Commission first included satellite Internet in the scope of "new infrastructure." Since implementing this plan, regions like Beijing, Shanghai, Guangdong, Sichuan, and Hunan have issued relevant industrial policies to support the development of space-based information industries like satellite communication. In January 2021, Beijing issued Several Measures to Support the Development of the Satellite Network Industry, proposing 26 tasks and 3 work protection measures in 8 areas, such as creating a new highland for scientific and technological innovation.

In March 2021, China's 14th Five-Year Plan and Long-Range Objectives through the Year 2035 reiterated the goal of building a high-speed, ubiquitous, integrated, and secure information infrastructure. By 2025, China's satellite communication network is expected to provide global information network services for various land, sea, air, and space users.

Furthermore, policy support has driven the accelerated development of the satellite communication industry. Since 2016, China has initiated multiple near-Earth satellite constellation plans, including projects like Hongyun, Xingyun, Hongyan, and Tianxiang. The Hongyun Project, proposed by CASIC, plans to launch 156 satellites for global networking. According to the plan, the entire Hongyun Project is divided into three steps: the first step involves launching a technical verification satellite by 2018, which has been completed; the second step involves launching four experimental satellites by the end of the 13th Five-Year Plan period to form a small constellation; the third step is to realize the networking of all 156 satellites by the end of the 14th Five-Year Plan, completing the construction of the service constellation. The Hongyan constellation plans to launch 300 LEO communication satellites to establish China's first space communication network that combines wide and narrow bands.

In April 2021, China Satellite Network Group Co., Ltd. ("China StarNet") was established to overcome key technological and frequency-orbit resource bottlenecks in constructing LEO satellite Internet. China StarNet, the first central enterprise registered in Xiong'an New Area, is the only state-owned backbone enterprise approved by the central government to design, construct, and operate satellite Internet.

The establishment of China StarNet drew high attention from all sectors of society, and many believe that its establishment aims to coordinate national and private capital to promote the maturity of the satellite communication industry chain, marking the acceleration phase of China's satellite Internet construction. *Global Magazine* wrote, "This is the satellite Internet appearing again in the public view as a major national strategy of China since it was included in the new infrastructure in 2020."

Notably, a year before the establishment of China StarNet, in September 2020, a Chinese company codenamed "GW" submitted a spectrum allocation dossier to the ITU. The dossier revealed two broadband constellation plans named GW-A59 and GW-2, planning to gradually launch 12,992 satellites to build a star network. The identity of "GW" was not clear then, and it did not attract much attention. Some media speculated that "GW" might be related to Galaxy Aerospace's constellation plan. However, Galaxy Aerospace had already announced its "Galaxy Series" constellation plan of 650 satellites. It wasn't until the establishment of China StarNet that the mystery was unveiled. "GW" stands for the initials of "Guowang" in Pinyin, likely the provisional name before the establishment of China StarNet.

The "StarNet" constellation shares many similarities with "Starlink," such as the astonishing number of satellites, orbit altitudes divided into two groups (one in extremely low orbit and one in near-Earth orbit), and orbital inclinations ranging between 30°–85°. Both are global satellite communication networks. Although China had earlier constellation plans like "Xingyun," "Hongyun," and "Hongyan," "Xingyun" was mainly for narrowband IoT, while "Hongyun" and "Hongyan" were still in the early stages, and their overall constellation scale was somewhat insufficient compared to similar foreign satellite constellations.

Earth's near-Earth orbit is estimated to accommodate only about 60,000 satellites. With SpaceX already planning 42,000 satellites, occupying a substantial portion of the Earth's very low and near-Earth orbits, the addition of China's planned 12,992 satellites, combined with other major constellation plans globally, means that Earth's low orbit is already overloaded. Therefore, the "StarNet" constellation plan also plays a role in competing for orbital resources, preventing China from facing a situation with no available orbits in the future. From this perspective, the plan is significant, and the earlier the launch, the better.

Currently, "StarNet" satellite tenders and launch base construction have been completed sequentially, and the first satellite launch is imminent. On October 18, 2022, China StarNet Network System Research Institute Co., Ltd. announced the winning bidders for the communication satellite, including the China Academy of Space Technology, Shanghai Microsatellite Engineering Center, CETC 54, and Galaxy Aerospace. The StarNet satellite launch base, located in Dongjiao Town, Wenchang City, Hainan, officially started construction in July 2022, and its main structure has been capped. The first launch position is expected to be completed by the end of 2023, with the first satellite scheduled for launch in the first half of 2024.

Meanwhile, on July 9, 2023, China successfully launched a satellite Internet technology test satellite into the intended orbit from the Jiuquan Satellite Launch Center using a Long March 2C rocket. Additionally, in June 2023, at the second working group meeting of the ITU Radiocommunication Bureau's Satellite Research Group, a proposal led by China's Academy of Information and Communications Technology, including CITIC Mobile, Shanghai Microsatellite Engineering Center, and others, was officially passed. The proposal covered key technological directions like direct-to-mobile satellite communication, onboard processing, inter-satellite links, high-low orbit satellite collaboration, and spectrum sharing between satellites and terrestrial technologies, marking significant progress in developing integrated space and terrestrial technology standards. With "StarNet" driving the initiative, China's launch of StarNet satellites is expected to enter the fast lane.

However, compared to Starlink, China's StarNet may face some constraints:

1. **High satellite manufacturing costs.** Thanks to assembly line production and mass manufacturing, the cost per Starlink satellite can be less than $1 million, with Elon Musk himself stating that it could eventually be reduced to $500,000. In contrast, China has primarily launched medium and high-orbit satellites in the past, focusing more on reliability due to the need for fewer satellites for global coverage. Traditional satellite production methods involve fixed station production, with several engineers completing thousands of processes and extensively testing components, assembly, and pre-launch. Due to the lack of industrial clustering, some satellites even require remote testing. Therefore, assembling and launching a single satellite takes eight to ten months, costing around CNY 30 million per satellite. Although future manufacturing costs are expected to decrease, they remain higher than Starlink's. It's worth noting that with the advancement of domestic constellation plans, the G60 Starlink plan, jointly released by nine cities in the Yangtze River Delta's "G60 Science and Innovation Corridor" with Shanghai's Songjiang district leading, is building the first satellite manufacturing "Lighthouse Factory" in the Yangtze River Delta. This initiative aims to accelerate the gathering of upstream and downstream enterprises in the industry chain, creating China's first satellite Internet industry cluster. Once operational, the factory's capacity is expected to reach 300 satellites/year, reducing the production cycle by 80% and the cost per satellite by 35%.

2. **Rocket launch costs and technological advances.** China faces a significant challenge with higher rocket launch costs than SpaceX's rocket capacity advancements, multi-satellite launch capabilities, and rocket reusability. SpaceX has reduced costs to as low as $1,520/kg, while China's costs are still several times higher. To compete, China must enhance single-rocket capacity and master rocket recovery technology. Efforts are underway, with the China Aerospace Science and Technology Corporation's Long March 8 rocket team working on key recovery technologies. Private enterprises like Deep Blue Aero-

space and iSpace are also making strides, with successful low-altitude vertical takeoff and landing tests and plans for orbital launch and recovery missions.

3. **Market access and policy optimization**. China is progressively liberalizing market access and enhancing regulatory frameworks. This strategic approach aims to create cost-effective manufacturing and launching solutions and to develop robust market demand for operational applications. While replicating SpaceX's model might be challenging, these advancements could establish a sustainable and prosperous path for China in satellite communications.

4. **Private sector initiatives**. Galaxy Aerospace has reported plans for over 1,800 satellites, including successfully launching the world's first Q/V band low-orbit broadband satellite. Geely's future constellation plans to deploy 240 satellites by 2025.

5. **Beyond LEO**. China also invests in higher and medium orbits. China Satcom's "Hai Xingtong" already serves over 6,000 vessels, and China Telecom's satellite communication division, "Tiantong-1," has launched its first satellites, covering China and its maritime regions.

6. **Ultra-low orbit communication and sensing constellation**. In late 2023, China began developing an ultra-low orbit communication and remote sensing constellation. With plans to launch more than 3,900 satellites by 2027 and over 6,000 by 2030, China is set to be a formidable force in the global satellite communications market.

7. **Global competition**. Beyond the US, Europe, Russia, and China, Canada's Telesat, and South Korea's Samsung are also planning large-scale satellite constellations. Telesat's "Lightspeed" project aims to launch 198 satellites by 2027, while South Korea plans to develop a constellation of 100 mini-satellites over a decade. Japan is collaborating with corporations and universities to develop a new generation of satellite communication networks using small satellites.

8. **The non-renewable nature of spectrum and orbits**. Given the non-renewable nature of spectrum and orbits and the ITU's policy of "first come, first served," nations worldwide are vigorously supporting their aerospace enterprises to gain a foothold in the satellite Internet field. China's comprehensive approach and strategic initiatives make it an essential player in this global competition.

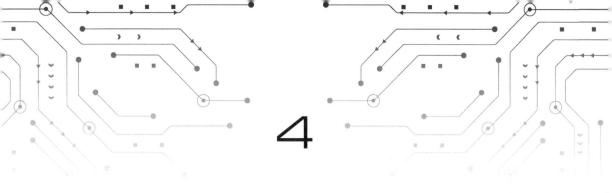

4

DISMANTLING THE SATELLITE COMMUNICATIONS INDUSTRY CHAIN

4.1 Upstream: From Manufacturing to Launching Satellites

Satellite communication is a complex industry. Looking at the industrial chain of the satellite communication sector, the upstream part mainly involves satellite manufacturing and launching. The midstream includes the manufacturing of satellite ground equipment and satellite communication operations, while the downstream is concerned with ground terminals and satellite communication applications. From satellite manufacturing to launching and ground facility manufacturing to satellite communication operations and services, each segment is a complete industrial chain. Let's first look at the upstream part.

4.1.1 Satellite Platform, the Starting Point of Satellite Manufacturing

To launch a communication satellite, the satellite itself must first be manufactured. Regarding satellite manufacturing, a communication satellite generally consists of two parts: the satellite platform and the communication payload.

The satellite platform, comprising the satellite's main body and support systems, is akin to a "chassis" for the satellite, capable of accommodating different functions and applications. For example, trucks are usually sold as chassis without specific customizations. Buyers often purchase either the chassis or pre-fitted cargo boxes from dealers. Those who buy just the chassis can customize it to their needs, which we call "upper body" modifications. This can include fitting cargo bays, tankers, dump beds, ladders, or even cranes, transforming the truck into a van, flatbed, high-sided truck, refrigerated truck, etc. Some are even modified into specialized or recreational vehicles. Similarly, the "satellite platform" provides the basic functionality for a satellite, which can then be fitted with various specialized modules depending on the mission.

Regardless of the payload installed, most satellites have some consistent basic functions, though specific technical performance may vary. Many countries worldwide adopt this truck chassis-like approach to "common satellite platforms" in satellite design and development. They design universally applicable satellite platforms whose interfaces and power parameters can adapt to different payloads within a certain range. When loading various payloads, the satellite platform only requires minor adaptability modifications.

Essentially, the fundamental function of the satellite platform is to ensure the normal operation of the satellite. It includes multiple systems, such as the energy system, attitude and orbit control system, propulsion system, telemetry and remote-control system, and thermal control system.

The energy subsystem provides power for the entire satellite. The attitude and orbit control subsystem, equipped with various sensors, primarily ensures that the satellite's attitude orientation and orbit positioning errors remain within allowable limits. The propulsion system provides the momentum for satellite orbit insertion, orbit maintenance, and attitude control. After a satellite is launched into its orbit, the main factor influencing its lifespan is not the lifespan of electronic devices, but the amount of fuel available for maintaining its stationary orbit. The thermal control system ensures that various devices on the satellite operate at appropriate temperatures. The telemetry, tracking, and command systems maintain contact with the ground control center.

Overall, the core role of the satellite platform is to provide maneuverability and power for the satellite, and the related subsystems also constitute the largest cost proportion. According to data from iResearch, a consulting firm in China, the cost structure of the satellite platform shows that the attitude control system (including the propulsion system) accounts for 40% of the costs, and the power system accounts for 22%, totaling over 60% and representing the largest cost components of the satellite platform.

4.1.2 The Payload of Communication Satellites

Another crucial component of satellite manufacturing besides the satellite platform is the satellite payload. The payload is essentially the "tool" the satellite uses for its operations in space. For instance, the primary payload of observation satellites is usually a camera. Depending on their application, satellite payloads can be categorized for communication, weather, Earth resources, marine, navigation, surveillance, scientific, and technology test satellites.

For communication satellites, their payload, in conjunction with ground satellite communication equipment, forms the satellite communication system, handling the reception, conversion, amplification, and transmission of signals from Earth stations.

The payload of communication satellites differs significantly from ground communication equipment. It requires radiation-resistant components, high demands for miniaturization and low power consumption, and the capability to operate continuously and reliably over extended periods. The payload of military communication satellites should also be able to resist intentional interference and ensure the effective use of various small, mobile (moving) Earth stations over a large geographical area.

The payload of communication satellites mainly consists of onboard antennae and transponders. The traditional payload of communication satellites includes onboard antennae and transponders, whereas satellite Internet systems add inter-satellite links to the traditional communication satellite payload.

4.1.2.1 Antennae: The "Eyes" and "Ears" of Satellite Systems

Onboard antennae are the "eyes" and "ears" of satellite systems, responsible for converting electromagnetic signals in space into output or input signals for the satellite. They play a crucial role in receiving and transmitting information, akin to humans' visual and auditory systems. Additionally, the value proportion of onboard antennae in satellite manufacturing is significantly high. According to iResearch, the platform and payload each account for 50% of the value of custom-made satellites. In contrast, in mass-produced satellites, the payload can constitute up to 70% of the value, with the antenna subsystem being the core part of the communication satellite payload.

Currently, the primary type of onboard antenna is the multi-beam antenna, a system capable of simultaneously pointing to or covering multiple directions. Unlike traditional single-beam antennae, multi-beam antennae can simultaneously process and transmit multiple signals, allowing communication or observation with multiple targets. Multi-beam antennae are classified into three types based on their form: multi-beam reflector antennae, multi-beam lens antennae, and multi-beam phased array antennae, with the latter becoming the standard for LEO communications.

In the past, antennae, similar to signal-receiving "dishes," found it challenging to maintain continuous connections with LEO satellites due to their rapid orbital movement. Tracking these fast-moving satellites and swiftly switching from one to another without causing communication interruptions was difficult with a single antenna. Multi-beam phased array antennae, however, change the shape of the directional pattern by controlling the feeding phase of the radiating elements in the array antenna. Compared to traditional antennae, phased array antennae do not require mechanical rotation. They adjust the signal direction to align with satellites through remote control of signal phases and can support multiple satellites simultaneously, greatly benefiting mid and low-orbit satellite communications.

In 1987, Motorola proposed and deployed the first genuine global personal communication system reliant on a LEO satellite constellation named Iridium. The Iridium constellation comprises 66 satellites distributed across six polar orbital planes, with an orbital height of 780 km and an inclination of 86.4°. The primary mission antennae of each Iridium satellite comprise three L-band active multi-beam phased array antennae used for sending and receiving, angled toward Earth to provide L-band links from the satellite to ground users. The US IridiumNext constellation equips each satellite with three active phased array antennae, each capable of generating 16 beams. The rapid deployment of the Starlink project today also owes much to phased array technology.

Specifically, a phased array antenna consists of multiple units, each independently functional and arranged according to certain rules. Each radiating element's feeding phase and amplitude can be flexibly controlled, forming a large array to achieve greater antenna gain, thereby enabling rapid network access and high-speed data communication.

Based on the architecture of phased array antennae, they can be divided into passive electronically scanned arrays (PESA) and active electronically scanned arrays (AESA). PESA has only one central transmitter and receiver. The high-frequency energy generated by the transmitter, actively distributed by a computer to each unit of the antenna array, produces spherical wavefronts from each antenna radiator. These wavefronts add up in front of the antenna to create a plane wave traveling in a specific direction. By controlling the phase shift through a computer, the beam's pointing angle θ is altered. In AESA, each antenna unit is equipped with a transmit/receive (T/R) module, with each T/R module capable of independently emitting and receiving electromagnetic waves. AESA has advantages in bandwidth, power, efficiency, and redundancy design, and its fault-tolerance feature makes it more suitable for satellite applications.

The T/R module is the core of the phased array antenna. According to the Research on Low-Cost Active Phased Array Antennae, the cost of T/R modules accounts for 50%–60% of the cost of an active phased array antenna module. In operating a phased array antenna, the T/R module is responsible for transmitting and receiving signals and controlling their amplitude and phase, thereby completing beam shaping and beam scanning, with its indicators directly affecting antenna performance.

Thanks to advancements in integrated chip technology, phased array antennae can achieve extremely high integration, leading to lightweight antennae. Globally, the US and Germany are leaders in phased array technology. Although China started relatively late, its development has been rapid. Currently, China is accelerating the layout of satellite Internet, and demand-driven development is prompting phased array antenna technology iteration. For instance, Galaxy Space independently developed China's first Q/V band communication capacity LEO broadband communication satellite, accelerating the demonstration of the next-generation broadband digital processing payload, and will apply phased array antenna technology to achieve flexible scanning and coverage of LEO satellite user beams.

As phased array products become more lightweight and cost-effective, they will unleash tremendous value in various military and civilian fields, such as satellites, unmanned aerial vehicles, ground communications, and even autonomous driving. In the future, phased array antennae could even integrate into our daily lives. For example, car roofs might embed phased array antennae to communicate with satellites anytime, significantly promoting the realization of truly intelligent driving, or our clothing could become computer screens, with such garments potentially embedding phased array antennae for signal transmission and detection, offering a viable solution for realizing the metaverse.

4.1.2.2 Transponders: Facilitating Signal Relay and Forwarding

Onboard transponders are responsible for transmitting and forwarding signals in satellite communications, serving as the core module in onboard signal processing and exchange technologies and key technology in satellite communication networks.

Two types of transponders correspond to transparent and regenerative payloads: transparent transponders and regenerative transponders.

Transparent transponders primarily consist of high-power amplifiers. They have a large capacity and a simple structure, including complete uplink and downlink pathways. The terminal determines band division, but they have weaker anti-interference capabilities. Information transmission between terminals requires two hops.

Regenerative transponders employ regenerative onboard processing technology. This involves demodulating and decoding all user signals that require processing, followed by re-encoding and demodulating after switching. Compared to transparent transponders, regenerative transponders have separate uplink and downlink pathways. By demodulating and decoding, they eliminate noise accumulation and offer strong anti-interference capabilities, higher spectral efficiency, and improved communication quality. Notable regenerative transponders include North America's SpaceMux, Europe's Skyplex, and Japan's WINDS.

Transponders consist of analog circuits and digital processing chips. Specifically, onboard transponders receive uplink signals from ground stations, which, after input filtering, are broadband amplified by the low noise amplifier in the receiver. The signal frequency is then converted to a downlink signal using the frequency converter in the receiver, followed by branch filtering for channel control and amplification of the signal power using one or more power amplifiers. Finally, power synthesis is performed using an output multiplexer. The combined downlink signal is sent back to the ground via the transmitting antenna, completing the signal relay and forwarding task. This process essentially organizes, amplifies, and adapts the information sent from ground stations for space transmission.

Power amplifiers are key components of onboard transponders. Their energy conversion efficiency directly impacts onboard thermal processing and payload capacity. Currently, the high-power amplifiers used mainly include three types: Traveling Wave Tube Amplifier (TWTA), Solid State Power Amplifier (SSPA), and Klystron Power Amplifier (KPA).

TWTAs use thermionic emission, requiring precise high-voltage power supplies, thus having high production and maintenance costs. KPAs have a bandwidth of only 50–100 MHz. With the continuous development of solid-state technology and power-combining technology, SSPAs have gradually achieved high power output. SSPAs are being increasingly accepted in the design of LEO satellite communication system payloads due to their good linearity, as exemplified by the first experimental satellite launched by Galaxy Space, which utilized an SSPA. The era of satellite communication will create a broader market prospect for transponders.

4.1.2.3 Laser Communication: The Core of Inter-satellite Links

The networking of communication satellites involves three links: Inter-Satellite Link (ISL), Service Link, and Feeder Link. ISLs are the links between satellites, Service Links are between satellites and users, and Feeder Links are between satellites and teleports. Of these, ISLs are unique to satellite Internet. Employing ISL supports establishing a global satellite backbone network, enabling global satellite management and control. This significantly enhances independence from ground systems, expands system communication capacity, and overcomes geographical limitations in ground station data transmission, thus improving the system's resilience, autonomy, mobility, and flexibility.

Satellite communication methods are mainly divided into two types: using electromagnetic waves and using light. Further subdivisions include microwave, terahertz, laser, and quantum communication. For ISLs, traditional satellites usually rely on microwave/millimeter-wave communication technology. However, with the growing global communication demand, the limitations of traditional satellite communication technology using microwaves are becoming apparent. First, long-distance microwave communication requires high power consumption and has limited transmission rates. Second, microwave communication requires specific frequency bands to avoid overlapping with neighboring satellite frequencies and prevent signal interference due to the complex and variable interstellar environment.

Microwave communication needs help to meet future high-speed, broadband communication needs. Therefore, ISL technology must evolve toward smaller, lighter, and lower power consumption to support the increasing data transmission demands. In this context, laser communication technology is rapidly developing and replacing traditional microwave communication technology. Inter-satellite laser communication terminals are opto-mechatronic integrated systems. Establishing inter-satellite laser links primarily depends on capturing, tracking, and pointing. In 2021, the US Space Development Agency launched four satellites to validate key technologies for the "Next-Generation Space Architecture," including inter-satellite and satellite-to-unmanned aerial vehicle laser communication technologies.

Satellite laser communication is categorized into two types based on the transmission environment: laser communication in a vacuum environment, i.e., inter-satellite laser communication, mainly for devices in a vacuum, such as communication between satellites, spacecraft, space stations, etc., and laser communication in an atmospheric environment, i.e., satellite-to-ground laser communication, which is widely used for connections between satellites and ground, maritime users, and aircraft. Core technological elements of satellite laser communication include key components, communication systems, and alignment and capture methods. Key components include laser transmitters, transmission optical lenses, receiving optical lenses, laser receivers, etc.

From a technological perspective, precision tracking and low-cost design are our current research and development focus; medium-term goals include solving atmospheric correction models and higher speed, long-distance communication technologies for satellite-to-ground and ground-to-ground communications. Looking further ahead, the integration and networking of laser communication will be the focus of technological breakthroughs.

Apart from laser communication, terahertz and quantum communications also receive significant attention. However, these technologies or related devices still need to be mature enough for industrial use and are still distant from practical applications.

Satellite laser communication has garnered considerable attention because it offers clear advantages over traditional satellite microwave communication technology, making it the core method for establishing ISLs.

First, laser communication has unrestricted channel resources. The frequency bandwidth available for laser communication reaches GHz levels, and inter-satellite laser communication does not require specific frequency band applications to the ITU, making channel usage more convenient. Second, laser communication has a high communication rate. The frequency of lasers is three to four orders of magnitude higher than microwaves, with communication rates reaching 10 Gbit/s or higher, allowing for the transmission of large amounts of data in a short time. Third, laser communication has low power consumption and strong anti-interference capabilities. Laser beams have excellent directivity, concentrating energy during transmission, making them less likely to disperse. This is advantageous when high link communication rates are required, as the benefits of lasers in terms of size, weight, and power consumption become evident. The extremely narrow beam divergence angle also makes the signal difficult to intercept and interfere with. Inter-satellite communication channels are also free from atmospheric interference, making them highly suitable for developing laser communication. Currently, due to its high channel throughput, high transmission bandwidth, strong anti-interference capabilities, and high confidentiality and security, laser communication has become the optimal solution for establishing ISLs.

Presently, foreign communication, remote sensing, navigation, and relay satellite systems have all planned to deploy laser communication terminals. Laser inter-satellite link technology has evolved from in-orbit demonstration to large-scale networking applications. Multiple mainstream LEO constellation systems have proposed developing ISL capabilities, with Starlink making laser inter-satellite links one of its core transmission link methods. From 2022, its second-generation satellites have been equipped with laser communication terminals, making them a standard payload for Starlink satellites.

China's satellite laser communication is also transitioning from the experimental verification stage to the productization stage, showing a trend of accelerated development. On August 13, 2020, China successfully verified laser communication payload technology aboard the "Xingyun-2" twin satellites, achieving a breakthrough in satellite IoT constellations for inter-satellite laser communication.

In China, Shangguang Communication is a leader in satellite Internet communication terminals. Originating from the Chinese Academy of Sciences' Key Laboratory of Space Laser Information Transmission and Detection Technology, Shangguang Communication is one of the earliest units in China to start space laser communication research, holding significant technological advantages and ten years of engineering experience. As one of the first commercial companies established in this field, Shangguang Communication received orders for inter-satellite laser communication terminals for dozens of satellites in multiple projects in the first phase of the StarNet bid. Backed by the Shanghai Institute of Optics and Fine Mechanics, Shangguang Communication has accumulated delivery and verification experience for over ten sets of terminal products. It is achieving an annual production capacity of a hundred sets and aims to reach 300 sets by 2024, providing a stable equipment supply for StarNet construction. Shangguang Communication is accelerating its commercialization progress by leveraging scale advantages to reduce costs. Given the planned number of LEO satellites, the laser communication terminal market is expected to reach a scale of hundreds of billions in the mid-term, with Shangguang Communication poised to become a leading company in this field.

4.1.3 Decreasing Costs in Satellite Manufacturing

The manufacturing cost of communication satellites is a crucial factor affecting the commercial viability of the satellite communication industry. Currently, communication satellites' research and development costs remain high due to production scale limitations. However, with the recent surge in demand for satellite constellations by overseas satellite network companies like OneWeb and Starlink, the mass production of LEO communication satellites has continuously reduced manufacturing costs.

According to public information, in 2022, OneWeb could produce one to three satellites per day, with the cost of a single satellite around $600,000. Starlink's production capacity reached seven to eight satellites daily, and media reports suggest that the cost of manufacturing a single Starlink satellite has dropped to approximately $300,000. This cost reduction is attributed to Starlink's mature supply chain and large-scale production. Mass production also allows Starlink to distribute research and development costs effectively and continuously refine its supply chain, reducing costs from multiple aspects.

In China, public information about LEO communication satellites is scarce. Based on public data from Galaxy Space, the Chinese Electronics Enthusiasts Network estimated that in 2022, the development cost of a single second-generation LEO broadband communication satellite from the company was between CNY 10 million and 15 million. The weight of these satellites is similar to the first generation of Starlink satellites, but their communication capacity is more than double, giving them a performance advantage.

Several Chinese companies possess satellite platform solutions and mass manufacturing capabilities regarding satellite assembly. These include the CASC, CASIC, Galaxy Space, Geespace, MINO SPACE, COMMSAT, Chang Guang Satellite, ZeroG Lab, etc. Private commercial satellite companies like Galaxy Space and GEESPACE are developing rapidly and have successfully launched mass-produced satellites into orbit.

However, Western companies appear more mature in cost control. For instance, OneWeb adopted an industrialized, automated, and standardized production line similar to aircraft manufacturing, using assembly line methods to increase production efficiency and reduce costs. Starlink, through a high proportion of self-developed components, including inter-satellite laser communication equipment, Hall thrusters, specialized chips, and Field-Programmable Gate Arrays (FPGA), and by opting for more affordable industrial or consumer-grade chips instead of expensive "aerospace-grade" chips, has significantly compressed costs.

According to public information, if we assume that the platform costs account for 30% and the payload costs 70% of a communication satellite's expenses, the antenna subsystem constitutes 75% of the payload cost. The antenna subsystem represents over 50% of the total satellite cost. Within this, the cost of T/R components accounts for about 50% of the antenna subsystem. Therefore, T/R components are indispensable when discussing communication satellite components.

The trend for T/R components is integrating multiple devices onto a single Monolithic Microwave Integrated Circuit (MMIC). Additionally, T/R components are transitioning from GaAs to GaN materials. As a third-generation semiconductor, GaN offers the advantage of a wide bandgap. It performs better than GaAs in RF applications, providing higher peak power, lower cost, and greater efficiency for the same volume as GaN MMIC.

Furthermore, FPGAs, crucial for processing high-speed digital signals, play an important role in communication satellites. However, due to the presence of high-energy particles and cosmic rays in space, which can damage components or cause "single-event upsets," FPGAs used in communication satellites must have radiation resistance, i.e., be "aerospace-grade."

Western companies like Xilinx, Actel, and Microchip (which acquired Atmel) offer several widely used aerospace-grade FPGA products. Chinese academic papers have revealed that China's Chang'e 4 lunar lander used CPUs and FPGAs from Atmel. In China, organizations like the 771st Research Institute of CASC (Xi'an Microelectronics Technology Institute), the 772nd Research Institute (Beijing Microelectronics Technology Institute), and Zhuhai Orbita are leading in aerospace-grade FPGA development. For instance, some products from the 772nd Institute have been used in the BeiDou-3 satellites and the Shenzhou-14 spacecraft.

GEO communication satellites are much more expensive than LEO communication satellites. They require tremendous communication capacity and are significantly larger and heavier than LEO communication satellites. Their satellite platforms and payloads are also more complex. Consequently, GEO communication satellites have a lower overall demand, and their research and development costs remain high. For example, ViaSat's communication satellites in GEO are publicly priced at approximately $360 million per satellite.

4.1.4 Rockets: The Core of Satellite Launches

Following the manufacturing of satellites, the next step is launching them into space. What launches satellites into space? Rockets. Without rockets, satellites would remain mere Earthbound objects. Therefore, rockets are the focal point of the satellite launching phase and comprise most of the satellite launch costs. As a result, in the satellite launch phase, the commercial satellite industry focuses primarily on the cost of rockets.

As a system engineering project, rocket development can generally be divided into three main stages: design, production, and testing. According to the 2021 China Commercial Aerospace Industry Development Report by iResearch, research and development expenses in the design phase account for 70% of the initial research funds for a new rocket model, while production and testing account for 30%.

In the design phase, each new rocket model requires a redesign for different launch missions, making this the most optimization-needing business phase. Design and production are generally combined. After the mass production of rockets, the costs of these two phases can be reduced.

Looking at production, the cost structure of mainstream liquid rockets mainly includes four parts: propulsion system, electrical system, structure, and ground system. The propulsion system accounts for 70% of the total rocket cost, followed by the electrical system at 15%, the structure at about 8%, and the ground system at about 7%.

In this regard, the primary direction for cost reduction is clearly the propulsion system, with the engine being the key component. Engines are categorized into solid and liquid engines, with liquid engines having a clear advantage in achieving rocket reusability, making them the "heart" of reusable rockets. Presently, companies, including SpaceX and LandSpace, both domestically and internationally, focus their research and development primarily on liquid engines.

Globally, SpaceX's Falcon series, which is reusable, has a unique advantage in launch pricing. Before SpaceX, rockets used to send space shuttles and satellites into orbit were designed for single use, with no attempts made to design reusable rockets. This was directly related to the past perception of satellites, as only a few were launched over years or even decades, and rockets were custom-made as needed.

Clearly, SpaceX broke past perceptions of rockets. Typically, the rocket's first stage (accounting for 70% of the total rocket cost) falls back to the Earth's surface after use, burning up and being destroyed in the atmosphere. Reusable rockets are an attempt to solve this problem. The development of reusable launch space systems began in the first half of the 20th century with the Silber-vogel project. After the war era, spaceflight underwent rapid development, and reusable launch systems were a major development theme in the early 1970s. This period also saw the birth of the largest project in the field of reusable space-craft—the STS Space Shuttle program, the first operational reusable spacecraft. Currently, due to technical and economic reasons, these space shuttles have been retired. Now, SpaceX is the leading company pursuing reusable rocket tech-nology and has successfully developed rockets capable of multiple launches, including the repeated use of the first stage of the Falcon 9 and Falcon Heavy rockets. The theoretical lowest cost of launching into LEO with the Falcon 9 and Falcon Heavy rockets is around CNY 10,000. However, this lowest cost calcula-tion is based on the cost of a single rocket divided by the theoretical maximum LEO capacity. Still, it is unlikely to reach maximum capacity in actual applica-tions, so the cost per kilogram of launch may vary greatly. Currently, the low-est-cost launcher in China is the Long March 3B, with the minimum cost of LEO transport at CNY 42,200.

It's worth mentioning that launch cost is not the sole criterion for evalu-ating a rocket. For commercial satellite companies, medium rockets carrying large satellites are often unnecessary. Therefore, they opt for lower-cost launch methods, such as "ride-sharing" launches where their satellites are co-launched with others. At the same time, such launch methods are relatively cheaper but also subject to constraints like launch scheduling.

Commercial satellite companies' choice of rockets is not solely based on launch cost. Compared to medium rockets with larger payload capacities, like the Falcon 9, these companies often prefer smaller rockets with relatively higher costs per kilogram. Although the cost per kilogram for a single launch may be higher for small rockets, commercial satellite companies value their more flexible launch schedules and shorter wait times, which positively impact the rapid deployment of satellites and the execution of business plans.

The choice of the rocket is not just about pursuing lower launch costs but a decision made based on a careful consideration of cost, flexibility, and mission requirements. While commercial satellite companies seek optimal solutions, they are also gradually driving the development of the small rocket market, providing more diversified and flexible choices for future commercial satellite launches.

4.2 Midstream: Unlocking Immense Value

In the satellite communication industry, the manufacturing of ground equipment and satellite communication operations in the midstream are the most valuable segments. According to the Satellite Industry Report published by the SIA in 2022, the global satellite industry's total revenue in 2021 reached $279 billion. This market encompasses satellite manufacturing, satellite launching, ground equipment manufacturing, and satellite operation and services, with their respective market shares at approximately 5%, 2%, 51%, and 42%, respectively. This indicates that in the maturing phase of the satellite industry, the midstream is experiencing immense value growth.

4.2.1 Manufacturing of Ground Equipment

Satellite communication ground equipment mainly consists of ground stations and terminal equipment.

Ground communication equipment in satellite communication systems, collectively called ground stations, includes all ground facilities that maintain the normal operation of satellites in orbit and support user-to-user communication through satellite transponders.

Ground stations can be classified into two types based on their use or function. The first type includes satellite tracking and control management stations, which monitor and manage satellites in orbit. This category includes Telemetry, Tracking, and Command stations, In-Orbit Testing stations, Earth Station Verification and Acceptance stations, and Communication System Monitoring stations. The second type is satellite application system ground stations, which form satellite communication networks with satellites and provide satellite communication services. These include central, regional, teleports, relay, and user stations.

Ground stations are further divided by installation methods and equipment scale into fixed stations, mobile stations (including shipborne, vehicle-mounted, airborne, etc.), and transportable stations that can be disassembled and relocated quickly. Evidently, the ground system, particularly in terms of terminal equipment and accompanying services, involves numerous participants and has a broad market space.

Ground communication systems, teleports, and tracking and control stations are gaining widespread attention. Teleports, as data center nodes in satellite communication Earth-space systems, are responsible for distributing and collecting satellite communication business data. Teleports can complete internal data exchange within the satellite communication network and data routing to external networks; they also have network management and operational control functions and are responsible for overall network resource scheduling, system equipment management, and user service management. To successfully use satellite Internet, users need to connect to the satellite using satellite terminals, and the data from this satellite must be relayed through teleports to terrestrial public networks.

Apart from teleports, tracking and control stations are also essential for ground infrastructure. Tracking and control are key factors in ensuring the normal operation of satellites in orbit, especially for LEO satellites with relatively limited lifespans. High-reliability tracking and control management are needed to ensure safe satellite operation in orbit and stable business development, maximizing satellite usage time and efficiently utilizing satellite value. Third-party commercial tracking and control companies will likely benefit from constructing LEO satellite constellations. On the one hand, the rapid increase in the number of satellites in orbit due to LEO satellite construction generates more tracking and control demand. On the other hand, the initial investment and operational costs of building one's own tracking and control stations can be significant, especially for private commercial satellite companies.

Looking at terminal equipment, terminals can be divided into fixed and mobile terminals. Fixed terminals vary depending on the type of satellite. For example, ground terminals for GEO satellites can be relatively simple and technically mature. However, fixed terminals for LEO satellites need to be equipped with servo tracking systems or phased array antennae to ensure stable data transmission.

Due to power limitations, current LEO satellite broadband applications still require fixed antenna terminals for access. In Starlink's ground fixed terminal StarLink Dish, a phased array antenna is equipped, with the number of antenna elements reaching 1,264, and the RF circuit includes 79 multi-channel beamformer chips and 632 RF front-end chips, equating to one RF channel corresponding to two antenna radiating elements.

Similarly, Amazon Kuiper's Ka-band phased array antenna and OneWeb's antennae for community public places and air passenger transport scenarios are examples. As the number of satellite Internet users grows, related fixed terminals are expected to experience explosive growth, driving demand for related phased array antenna components and T/R chips.

Beyond fixed terminals, consumer electronics satellite communication has also begun to take off. The most representative application is mobile phones. Since Huawei and Apple successively launched phones supporting satellite emergency communication, the popularity of satellite communication mobile terminals has rapidly surged. However, due to power and antenna size limitations, mobile terminals like phones still need help achieving satellite broadband connectivity, focusing mainly on emergency communication applications. For instance, Huawei's P60 series smartphones and Watch Ultimate smartwatch launched in March 2023 both support two-way BeiDou short messages, enabling sending and receiving information in no-signal conditions.

Specifically, Huawei and Apple equipped their Mate 50 series and iPhone 14 series smartphones with a special chipset for satellite connectivity and GPS coordinates. For this application to function effectively, it must be perfectly synchronized with the satellite, and the established connection must be maintained long enough to exchange with the smartphone. To this end, each smartphone integrates top-tier GPS receivers to compensate for the Doppler effect, calculate future satellite positions, and coordinate communication through cellular transceivers and specific front-end chipsets to receive and send signals on S/L bands.

Moreover, more mature satellite mobile communication systems have already become widespread, with portable satellite phone products on the market well developed. For instance, China's Tiantong satellite mobile communication system covers both military and civilian needs, providing communication services for oceans, remote areas, etc., supporting handheld terminals, broadband portable communication terminals, mobile vehicle-mounted communication terminals, and more.

Additionally, China's listed company KINGSIGNAL launched a Ku-band portable satellite communication terminal in 2020, customized for the Asia-Pacific 6D satellite. Compared to traditional fixed stations, portable satellite communication terminals, with reduced equipment weight and volume while maintaining high-quality communication signals, expand more usage scenarios for satellite communication. Besides, KINGSIGNAL has also launched a one-dimensional phased array of on-the-move electric lines, Tiantong mobile satellite phones, and other terminal products.

China's listed company, Meng Sheng Electronics, focuses on "on-the-move" satellite communication terminals, typically including satellite automatic tracking systems and satellite communication systems, mainly used in airborne, vehicle-mounted, and shipborne transportation.

HWA Create Company has launched a series of terminal products for the "Tiantong-1" communication satellite, including portable Tiantong broadband terminals, handheld satellite phones, narrowband data transmission terminals, and shipborne satellite terminals.

In April 2022, 3GPP (3rd Generation Partnership Project) completed the freeze of the 5G R17 standard, introducing new features for satellite communication and supporting 5G NR for non-terrestrial networks (NTN), allowing smartphones and IoT devices supporting eMTC and NB-IoT to achieve low-rate satellite communication. Currently, baseband chip manufacturers like MediaTek, Qualcomm, and Unisoc have completed related tests, and it's foreseeable that satellite emergency communication will soon become widespread on smartphones.

4.2.2 Satellite Communication Operations

As the satellite industry matures, operational services represent another major focus in the satellite communication industry value chain.

Satellite communication operational services can be divided into space and ground segment operations. Space segment operations include renting and selling satellite transponders, which involve leasing or selling owned or leased satellite transponder resources to satellite users for corresponding applications.

Ground segment operations cover satellite mobile communication, satellite fixed communication, and other communication services. Operators use legally obtained (owned or leased) satellite transponder resources to build corresponding satellite communication networks or systems, providing users with voice, data, multimedia communication, and other services.

Based on the operating model, satellite communication operators can be primarily divided into two types: operators with their own satellites and agents (renting satellites).

4.2.2.1 Operations of Owned Communication Satellites

Operators of owned satellite constellations in China primarily include China Satcom (ChinaStar and Asia-Pacific series constellations), Galaxy Space, Guodian Gaoke, etc. In 2021, China Satellite Network Group Corporation was established to integrate and oversee China's LEO satellite constellation plans, with an expected composition of up to 13,000 satellites, constructing a domestic satellite Internet system. Internationally, major players include Intelsat,

Eutelsat (merged with OneWeb), Inmarsat, SES Global, Viasat, Iridium Communications, SpaceX, Amazon, and others.

China Satcom, China's core operator of GEO communication satellites, currently provides satellite communication services, including TV broadcasting, maritime communication, airborne communication, emergency communication, and transponder rental. China Satcom owns both the ChinaStar and Asia-Pacific series constellations. The ChinaStar series originated from the Dongfanghong satellite and initially belonged to China Communications Satellite Co., Ltd., China's first satellite operator. During the telecom reform, the satellite business of the China Telecom Administration was spun off to form China Satcom (established in 2001, known as "old Satcom"). During the third telecom industry restructuring in 2008, its basic telecom business was merged into China Telecom, while other businesses were integrated into the CASC (the "new Satcom"). After the restructuring, CASC integrated its internal satellite operation businesses, injecting equity from Sino Satellite, China Broadcasting Satellite, Dongfang Satellite, Asia-Pacific Satellite, etc. The Sino series was later renamed the ChinaStar series, and the Asia-Pacific series was also integrated. The company has launched ChinaStar-16, ChinaStar-19, and ChinaStar-26 regarding high-throughput satellite layout.

China Satellite Network Group Company Limited, China's core operator of LEO satellite Internet, was established on April 26, 2021. The newly established China Satellite Network Group is under the State-owned Assets Supervision and Administration Commission of the State Council. China Electronics Corporation, CASIC, etc., jointly established it. It will integrate and coordinate China's original LEO satellite constellation plans, positioning close to the current three major operators. China Satellite Network has obtained the second type of basic telecommunications qualification and can operate in the domestic VSAT communication business.

Galaxy Space is China's leading private enterprise in constructing LEO constellations. Founded in 2018, the company plans to build a constellation of 1,000 LEO broadband communication satellites. In 2020, Galaxy Space launched its first satellite (Xingyun-3), adopting the Q/V band with a design capacity of 24 Gbps. It has established a complete experimental communication network system, including teleports, user terminals, ground stations, operation, and control systems, achieving a breakthrough from zero to one.

In March 2022, Galaxy Space launched six satellites of Batch 02 into orbit, building a small experimental constellation system ("Little Spider Web"). The Batch 02 satellites have a design communication capacity of over 40 Gbps per satellite, averaging about 190 kg in weight, and can provide uninterrupted, low-latency broadband communication services for about 30 minutes. In addition to existing Beijing and Sichuan teleports, ground facilities include newly built Nantong and Sanya mobile teleports, achieving communication coverage in densely populated areas and coastal regions in China.

Guodian Gaoke independently constructs the Tianqi constellation, holding UHF communication frequency licenses. Established in 2015 and completing equity financing in 2021, the company owns UHF band radio frequency and value-added telecommunications business licenses, building the LEO satellite IoT "Tianqi constellation" independently. The constellation is planned to consist of 38 satellites in a 900 km LEO, each weighing about 20–50 kg. Currently, 15 satellites are in orbit for the first phase of network construction, with plans to complete the 38-satellite constellation network by 2023. In terms of business model, Guodian Gaoke currently offers self-developed products, software, and system integration solutions, including terminal products, teleports, backend application software, system integration, etc. In the future, the company plans to develop toward traffic billing. Currently, Guodian Gaoke has landed application cases, providing monitoring services for power tower bases, hydrology, containers, agriculture, forest fire prevention, etc., for government and special departments in power, transportation, etc. For consumer customers, the company has launched Tianqi IoT emergency rescue terminals.

Globally, Starlink represents operators of owned satellite constellations. In 2015, SpaceX proposed a large-scale mega-constellation plan. In 2018, two prototype Starlink satellites were sent into their intended orbits, and in March and November of the same year, the FCC approved Starlink's first-generation LEO constellation and VLEO constellation plans, respectively. In 2019, Starlink launched its first batch of v0.9 satellites into their intended orbits, starting constellation construction. In 2020, Starlink began beta service in North America. By the end of May 2022, SpaceX claimed to provide Internet access services in 36 countries/regions worldwide and plans to expand services to Asia, Africa, and the Middle East in 2023.

Starlink is also a global leader in LEO satellite Internet. According to Starlink's plan submitted to the FCC, the total number of planned Gen1LEO satellites is 4,408, and the total number of planned VLEO satellites is 7,518. Gen2 plans for nearly 30,000 satellites. Since its first launch in May 2019 (excluding the TinTin experimental satellites), Starlink has launched over 5,000 satellites. As of December 2023, Starlink has surpassed one million users worldwide.

Simultaneously, companies like Viasat and SpaceX are trending in self-developing and self-manufacturing satellites. Viasat was the first to achieve payload self-development on high-orbit high-throughput satellites. SpaceX, leveraging self-development and self-manufacturing advantages, significantly reduced satellite manufacturing costs by improving production models in mass production. OneWeb previously collaborated with Airbus to establish a joint venture, OneWeb Satellites, responsible for satellite production.

4.2.2.2 Agency Operations of Communication Satellites

In China, companies like China Telecom, MARINESAT, and CITIC Satellite partner with other companies that own satellite constellation resources, providing services to the local market through agency or leasing arrangements.

China Telecom is the only one of the three major operators with satellite network resources. As mentioned earlier, during China's telecom reform, China Telecom divested China Satcom in 2009, which was then restructured into the CASC. Subsequently, China Telecom did not abandon the development and application of satellite communication. In 2009, China Telecom Group established China Telecom Satellite Communications Limited; in 2017, China Telecom Corporation Limited established its Satellite Communications Branch.

China Telecom exclusively undertakes the mobile communication business operations of the Tiantong satellite, also providing satellite mobile services, including Iridium, Thuraya (Eutelsat), and maritime services. On August 6, 2016, Tiantong-1 01, the project's first satellite, was launched into space. Currently, three satellites have been launched, covering China's entire territory and territorial waters, along the "Belt and Road" regions, and most of the Pacific and Indian Ocean areas. In May 2018, China Telecom launched its satellite phone service with the number segment 1740. Per the company's official WeChat, as of November 2022, the Tiantong satellite had reached 150,000 users.

MARINESAT primarily provides maritime satellite communication and other value-added services to commercial ships. MARINESAT has cooperated with various VSAT satellite resource providers, including China Satcom and Iridium, to offer daily office communication (telephony, email), high-speed multimedia communication services, and other communication services to commercial ships, as well as value-added services like remote video monitoring. The company plans further expansion into terrestrial businesses.

In recent years, MARINESAT's market share has been rapidly increasing. The company's satellite communication service revenue has continued to rise, and through establishing subsidiaries or operation centers in Singapore,

Greece, and Malaysia, it has accelerated its global market expansion. Benefiting from this, the company's market share has increased. According to Valor data, in 2017, the company ranked 14th in the global maritime satellite communication service market and rose to 11th place in 2019. As per company announcements, MARINESAT currently holds the top market share in China's commercial shipping market and has entered the global top ten.

Currently, the construction of satellite Internet is accelerating. As satellite constellations are progressively completed, the share of operations and services within the satellite communication industry chain is expected to increase slightly. In contrast, the overall scale of the operations and services market will continue to rise significantly.

4.3 Downstream: Satellite Communication Moving toward Applications

In the satellite communication industry's downstream sector, which encompasses the specific application scenarios of satellite communication, there is a shift from video services to broadband access, revealing vast market prospects.

4.3.1 From Video Services to Broadband Access

Based on end-user application scenarios, satellite communication applications include video services, mainly for television broadcasting and video distribution, and data services, primarily for broadband access. Television broadcasting and other video services represent the largest and most mature segment in satellite communication services, relying mainly on high-orbit broadband satellites.

However, with network video development based on 5G communication, television broadcasting services are experiencing negative impacts. The growth of Internet services like streaming media is challenging the demand for satellite television. Against this backdrop, the growth driver for satellite communication services is gradually shifting toward data services.

According to Euroconsult, within satellite communication service revenues, income generated from data services is expected to grow rapidly (projected to reach $51.4 billion by 2031), while video services, including satellite television broadcasting, are anticipated to shrink—expected to lose 26 million direct-to-home users and 8,500 satellite TV channels by 2031.

Simultaneously, broadband access represents the largest future application scenario for broadband satellites. Euroconsult data projects that by 2030, the capacity of global high-throughput satellites will grow at a compound annual growth rate of 28%, with consumer broadband access expected to account for nearly 60% of this global capacity growth. In terms of market size, according to SIA data, global satellite broadband revenue in 2020 was approximately $2.8 billion, a 7.7% increase year-over-year, with a projected CAGR exceeding 20% from 2021 to 2030.

Particularly, satellite broadband communication in global remote and rural areas shows immense application potential. Developed countries have a significantly higher broadband penetration rate compared to developing countries. According to ITU data, the average number of fixed broadband users per 100 people globally was 17 in 2021, whereas in regions like Africa, it was only 0.6, indicating a significant broadband coverage disparity. Moreover, within most countries, there are notable differences in broadband coverage between rural, remote areas, and cities. Even in developed regions like Europe and America, rural broadband coverage levels are far lower than in urban areas.

Taking the US as an example, according to FCC data, nearly 9.5% of rural residents need a broadband provider capable of offering more than 25 Mbps download and 3 Mbps upload speeds without considering satellite broadband providers. This figure is significantly higher than the 0.7% in urban areas. As speeds increase, the disparity in broadband coverage between rural and urban areas becomes more pronounced. In such situations, remote and rural areas are expected to drive further development of satellite communication data services.

SpaceX's standard residential service is one of the most representative cases of broadband access applications. Starlink users purchase the Starlink Dish to access Starlink services and build a home local area network through a router. The complete hardware equipment in the US region is priced at $599, with a monthly fee of $110, offering expected broadband access services with 50–200 Mbps download speed, 10–20 Mbps upload speed, and latency within 20–40 ms.

4.3.2 Broad Application Scenarios of Satellite Communication

From the perspective of specific application scenarios, satellite communication is a key piece in constructing an integrated network of space, air, land, and sea.

An integrated network of space, air, land, and sea is based on terrestrial networks and extends into space networks, covering natural spaces such as outer space, air, land, and sea. It provides information assurance for the activities of various users. Its goal is to expand the breadth and depth of communication coverage, deeply integrating traditional cellular networks with satellite communication (non-terrestrial communication) and deep-sea long-distance communication (underwater communication). Fundamentally, an integrated communication system of space, air, land, and sea consists of two subsystems: the integrated space-terrestrial subsystem combining terrestrial mobile communication networks with satellite communication networks, and the deep-sea long-distance (underwater communication) communication subsystem combining terrestrial mobile communication networks with deep-sea long-distance communication networks. As an essential part of the integrated information system of space and Earth, satellite communication is key to constructing an integrated network of space, air, land, and sea, helping to realize the blueprint for integration.

For instance, the International Maritime Satellite Organization uses communication satellites as relay stations for maritime wireless communication. The Inmarsat system consists of ship stations, shore stations, network coordination stations, and satellites. Its characteristics include high quality, large capacity, and global, all-weather, and full-time communication capabilities.

Due to the impossibility of establishing communication bases at sea, satellite communication is the only solution for ocean-going vessels. Satellite communication not only meets basic life needs such as entertainment and video calls for crew members with the development of the IoT and intelligent ship solutions, but it also connects ships in transit with shore-based data centers, achieving operational efficiency improvements and cost reductions through applications like energy optimization and condition monitoring.

According to the 2021 National Fisheries Economy Statistics Bulletin released by China's Ministry of Agriculture, by the end of 2021, China had 520,800 fishing vessels, including 357,000 motorized fishing vessels, most of which were long-distance fishing vessels. Additionally, with China's maritime law enforcement capabilities strengthening, the number of vessels required by maritime surveillance, coast guard, and other departments is expected to increase steadily. Limited by communication rates, tariff levels, usage habits, and other factors, satellite communication on ships has not yet been widely popularized, and the penetration rate is relatively low. With the advancement of high-throughput satellite technology, satellite communication charges will continue to decrease, promoting the formation of user habits. The shipborne market is expected to open up space, with the scale of China's shipborne satellite communication market projected to reach CNY 20.6 billion by 2025.

Aircraft communication provides Internet access services for airplanes, mainly through two technologies: the first is the Air-to-Ground (ATG) system, which receives signals through antennae installed on the underside of the aircraft and sends them to an onboard server. The modulated RF signal is then provided to passengers via Wi-Fi access points installed inside the aircraft. The GoGo ATG-4 system, for example, includes a main antenna on the underside, a side antennae, a server, and internal Wi-Fi access points. The second technology is satellite communication. The antenna that receives satellite signals is located on top of the aircraft and connects to the Internet via a satellite network. The GoGo satellite system, for example, includes a satellite antenna, an onboard server, Kandu (for controlling antenna movement), Modman (for signal conversion), and internal Wi-Fi access points.

Comparing the two technologies, ATG has lower speeds and is limited by ground network coverage, while satellite communication offers broader coverage and higher broadband application rates but higher equipment and maintenance costs. The development of low-orbit satellites can mitigate the disadvantage of high latency, and the cost per unit of bandwidth continues to decline, accelerating application.

In terms of market size, according to Euroconsult data, the global in-flight connectivity and entertainment applications market was about $2 billion in 2021, with the market size expected to exceed $5 billion by 2031. Currently, the development of aviation broadband in China could be much higher, with significant room for improvement. According to iResearch data, in 2020, domestic flight broadband coverage was about 13%, and international flight broadband coverage was about 37%. Some US airlines, like JetBlue, have achieved 100% in-flight broadband coverage.

Additionally, satellite communication can uniquely assist emergency rescue and disaster relief. In major natural disasters, terrestrial networks are usually paralyzed, making it difficult for external rescue commanders to grasp disaster situations in real time. Satellite communication, unrestricted by geographical coverage, can achieve rapid real-time connections, presenting disaster information to rescue commanders. It can also interconnect with terrestrial networks, further disseminating disaster information, gathering better rescue plans, and achieving emergency rescue.

Furthermore, satellite communication can play a unique role in emergency disaster relief. In major natural disasters, ground networks are often paralyzed, making it difficult for external rescue commanders to obtain real-time information on the affected areas, thereby hindering command and deployment efforts. Satellite communication, however, is wider than geography or coverage, enabling rapid and real-time connections. This allows disaster situations to be presented in real-time to rescue commanders, and it can also interconnect with ground networks, further disseminating disaster information and aiding in the development of better rescue plans, achieving emergency life-saving goals. Additionally, satellite communication can provide disaster warnings for extreme meteorological events such as river water levels, agricultural pests, forest fires, and Earthquake data.

Finally, satellite IoT will create new market entry points thanks to the expansion of application scenarios. From a market size perspective, according to ABI Research, by 2024, there will be 24 million devices accessing the IoT via satellites, and the resulting satellite IoT industry chain will be further refined and developed. NSR predicts that in the next decade, there will be two main types of terminals in the future space-based IoT: Mobile Satellite Communication System terminals and VSAT stations. These terminals are expected to grow at an annual rate of over 10%. Additionally, Asia will become the only region with a compound annual growth rate of over 10% in space-based IoT revenue. By 2027, Asia is expected to become one of the highest revenue-generating regions in the satellite IoT market, closing the gap with the North American market. With the gradual completion of various constellations, the market size for low-orbit small satellite IoT is expected to grow rapidly, from less than $20 million in 2020 to $130 million by 2027, with an average annual growth rate of nearly 70%.

Satellite communication based on the Starlink model is another major innovation in human informatization. Driven by this technological revolution, humanity is entering a new commercial era. For example, could future education evolve into a learn-as-you-go model, exploring nature while leveraging real-time online satellite communication systems and AI technology for teaching and learning? In healthcare, even an ordinary person without basic medical knowledge, when facing emergencies in the wilderness or environments without medical facilities, can receive emergency and professional treatment guidance from a real-time online AI doctor via satellite communication and emergency rescue calls.

The downstream of satellite communication is also undergoing a significant transformation from video services to broadband access, triggering a series of commercial changes. This transformation reflects the continuous development of technology, the increasingly diverse market demands, and the urgent need for high-speed data transmission. This change also demonstrates the vast market prospects of the satellite communication industry in broadband access and its profound impact on various industries. In the future, with continuous technological innovation and further diversification of market demands, satellite communication will continue to play a significant role in broadband access, providing more efficient and reliable communication services globally.

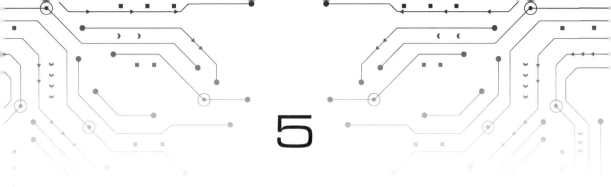

5

THE PAST, PRESENT, AND FUTURE OF COMMUNICATION

5.1   A Brief History of Human Communication

Since the dawn of humanity, communication has been a fundamental necessity. With it, people would be like isolated islands of information, able to collaborate, express emotions, or form any organization. In essence, communication is the glue of human society.

Of course, communication has taken various forms, and these have transformed significantly over time. From the ancient and sluggish beacon fire communication and pigeon post to today's 5G networks capable of supporting streaming media, human society has rapidly evolved with the advancement of communication technologies.

5.1.1   Ancient Communication: Arduous Journeys and Pigeon Post

Early humans had no language or writing; they were like animals, communicating only through bodily movements.

Several hundred thousand years ago, humans finally developed language. For a long time after that, communication was primarily face-to-face, with effective transmission limited to about a 50 m radius—the maximum distance a human voice could travel.

Between 3000 and 5000 BC, humans developed writing and drawing skills on top of spoken language. Writing overcame spoken words' temporal and spatial limitations, enabling ephemeral speech to be recorded and transmitted across distances and time. However, long-distance communication still needed to be more efficient, relying on manually delivered letters.

One famous story of this period is the origin of the Marathon: Pheidippides ran over 40 km to bring the news of victory to Athens. This story illustrates the early method of important state communication—by human running or walking.

To improve efficiency, humans began using tools for communication. The most well-known are beacon fires and pigeon posts. Beacon fire communication, originating in the Shang (1600–1046 BC) and Zhou (1046–256 BC) dynasties and lasting until the Ming and Qing (AD 1616–1911) dynasties, spanned several thousand years. Used mainly for military communication during wartime, beacon fires were built on high grounds and at strategic locations. When enemies invaded, guards would light fires or smoke signals as alarms. While efficient, this method had limitations, such as needing help to convey complex messages.

Pigeon post, another common ancient communication method, utilized pigeons' homing instincts, exceptional flying abilities, and remarkable memory. Important letters or intelligence were tied to pigeons' legs for transmission. Similar methods were used with other animals, like swans. However, pigeon post had its drawbacks: the risk of pigeons getting lost or intercepted and the vulnerability of the messages.

To address these issues, postal stations emerged, similar to modern courier services. Every city had a station where couriers on horseback relayed messages from one station to the next. These stations included land, water, and combined types, with the fastest horses covering over 150 km daily. By the Tang Dynasty, this system had become highly efficient, reaching its zenith in the Qing Dynasty with a network of 1,785 stations. This system remained until modern postal services replaced it in the mid-Qing period, but it was resource-intensive, exhausting both people and horses.

In the West, communication also has a long history. For example, in ancient Sparta, a Greek city-state, Spartans were among the first to use encoded messages (like the figure below). This required two specially crafted sticks, one for the sender and one for the receiver. The sender would wrap a piece of parchment around the stick, jumbling the letters. With the correct stick, the message is readable. The receiver would wrap the parchment around their matching stick to decipher the message.

Ancient Spartan communication

Subsequently, in the West, lighthouse communication became common, along with communication towers. The Chappe brothers in 18th-century France created a system of 16 towers between Paris and Lille, 230 km apart. Each tower had a mast with a movable crossbar, and messages were conveyed through different angles and positions of the crossbar, visible through telescopes. This method was particularly useful during the French Revolution.

Human communication has continually evolved toward greater efficiency and completeness, from beacon fires to sticks, pigeon posts, postal stations, and lighthouses. But there is still more to achieve.

5.1.2 Modern Communication: From Telegraph to Wireless Telephony

The advent of electromagnetic technology marked the transition from traditional to modern communication based on electromagnetic principles. In the early 19th century, through the relentless explorations of eminent scientists like Ørsted and Faraday, humans gradually unraveled the mysteries of electricity and magnetism. At that time, however, people did not realize that electricity and magnetism would offer a novel medium for information dissemination.

In 1832, inspired by electromagnetic induction, Samuel Morse resolved to invent a method of "transmitting messages through electricity"—the telegraph. Building upon the foundation of the "static electricity telegraph machine," Morse conceived his telegraph system, thus heralding the birth of telegraph technology. Using "electricity" for communication was a groundbreaking feat. Previously, despite thousands of years of experimentation with numerous communication methods, none could resolve the contradictions of message volume, communication delay, and distance. With its rapid transmission rate, electricity proved to be an ideal medium for communication.

Beyond the telegraph, Morse also invented the "Morse Code" to replace the 26 letters of the English alphabet in transmitting telegraphs. The significance of Morse Code lies in its advanced encoding of text.

We know that the birth of written language enabled true communication. Writing is essentially an early form of information encoding. All information can be recorded and transmitted through writing, forming the basis of human history and culture. However, writing has a drawback—its complexity. Even the 26-letter English alphabet needs to be more intricate for simplified transmission.

The secret to efficient communication lies in converting information into simpler codes and finding the most straightforward and efficient symbols to carry these codes. Morse Code is one such simplified encoding. It transformed text into "dots" and "dashes," achieving a minimalist "quasi-binary" encoding. Then, based on the "on" and "off" states of the electric current, it expressed these codes. This enabled transmitting text (information) over great distances at high speeds.

Utilizing the telegraph and Morse Code, Morse established a 64 km line between Baltimore and Washington. On May 24, 1844, Morse sent the first long-distance telegraph message from the Supreme Court Chamber in Washington to Baltimore: "What hath God wrought!" This invention of the wired telegraph rewrote human communication methods.

However, the telegraph had a limitation: the speed of generating electric pulses. Even a skilled telegraph operator took several seconds to tap out a word. In other words, Morse's telegraph could not facilitate instantaneous long-distance communication.

This limitation led to the invention of the telephone. Invented by Alexander Graham Bell in 1876, the telephone operated on a more direct principle than the telegraph. Bell did not encode messages but directly collected sound wave information through a diaphragm and magnet in the transmitter (microphone).

He then transmitted the electric current (mimicking the sound waveforms) to the receiver (earpiece), which drove a diaphragm and magnet to reproduce the sound waves. Functionally, the telephone was nearly the ultimate communication device of its time. Two people in different locations could converse directly. Its only drawback was the need for wiring, implying significant construction and cost.

Notably, almost simultaneously, wireless telephony began to develop. In 1888, the German scientist Heinrich Rudolf Hertz experimentally proved the existence of electromagnetic waves. In 1896, the Italian Guglielmo Marconi accomplished the first wireless radio communication over 30 m (later reaching 2 miles). Thus, humanity officially stepped into the era of wireless communication.

Although wired telephones and wireless telephony emerged around the same period, they developed independently for a long time.

For wired telephones, the primary challenge was converting voice signals into electrical signals, transmitting these via wires, and then converting them back into voice signals. The main issue in expanding wired telephony was the deployment and connection of these wires. With so many nodes to connect, switch processing was introduced. Initially, this was done manually, evolving into step-by-step and crossbar systems, forming a developmental trajectory for telephone switching.

However, whether step-by-step or crossbar, both used electromechanical actions for connection, categorizing them as "electromechanical automatic telephone exchange systems." But mechanical systems have their limits—low efficiency, small capacity, high failure rate—struggling to meet growing communication demands. Hence, there was a longing for a new switch processing method.

Finally, in December 1947, a research team from Bell Laboratories, comprising Shockley, Bardeen and Brattain invented the transistor. This innovation sparked the microelectronics revolution. With the rapid advancement of semiconductor and electronic technologies, the idea of incorporating electronic technology into telephone exchanges emerged. Initially, due to the inadequate performance of electronic components, a hybrid technology combining electronics and traditional mechanics, known as "semi-electronic" or "quasi-electronic" exchanges, was developed. Later, as microelectronics and digital circuit technology matured, the "fully electronic exchange" was realized.

In 1965, Bell Laboratories successfully produced the world's first commercial Stored Program Control exchange. In 1970, France inaugurated the world's first programmable digital exchange system, E10, in Lannion, marking the beginning of the digital exchange era. The essence of a programmable exchange is an electronically computed-controlled switch using pre-programmed operations for connections. Its advantages are evident: rapid connections, multiple functions, high efficiency, clear sound, reliable quality, and large capacity.

On the wireless communication front, the rapid development of integrated circuits and mastery of high-frequency electromagnetic waves eventually led to the miniaturization of wireless communication devices, paving the way for the birth of the mobile phone.

In April 1973, a man stood on the street of New York, holding a device about the size of two bricks, talking excitedly into it. This man was Martin Cooper, an engineer at Motorola and the inventor of the first true mobile phone, which could be carried by a single person and used for communication on the move.

The invention of the mobile phone marked the beginning of an era of universal communication and signaled the start of wireless communication, overtaking wired communication.

Interestingly, during the transition from wired landlines to wireless mobile phones, another communication method emerged—the pager. Known by various names such as beepers, pagers, BP machines, and BB machines, its professional name is "radio paging service." It was also widely known, especially in Hong Kong and Taiwan (China), by the nickname "Call Machine."

It's worth noting that a pager is not the same as a "big brother" phone or a "Little Smart" phone. It still required a mediator—for instance, if person A wanted to contact person B, they still had to use a public phone. A would tell the customer service operator B's pager number, and the operator would then page B's device, displaying A's public phone number. B would call back, allowing them to communicate finally. This was the analog phase of pagers, with relatively basic calling methods. The recipient had to dial the paging station to check their messages.

As pagers entered the digital phase, they began to support phone numbers and English message displays. Pagers could operate fully automatically at this stage, including directly checking the caller's number. Alternatively, customer service could convert a verbal message into a text sent to the recipient's pager, such as "Come home early" or "Call me back." The recipient would see the message and respond accordingly.

However, with the entry of mobile phones into the market, the pager business rapidly declined and eventually faded from people's lives. Today, pagers have become a relic of a bygone era.

5.1.3 Modern Communication: The Internet Revolutionizes Communication

The advent of the mobile phone initiated the era of modern mobile communication. The beginning of mobile communication is aptly termed the "1G Era" (see figure below for details). Motorola dominated this era, and the symbol of the 1G era was the "brick-like" large mobile phone. After 1980, these large phones gradually became a part of people's lives, enabling long-distance communication. However, 1G used analog communication technology with poor security, low capacity, subpar call quality, and unstable signals.

In the late 1980s, the transformation from analog to digital communication began with the maturation of large-scale integrated circuits, microprocessors, and digital signal-processing technology. Thus, the "2G Era" quickly dawned. In the 2G era, digital mobile communication technology made its debut. The transition from 1G to 2G represented a qualitative leap in communication capabilities—spectrum efficiency, anti-interference, security, and stability all improved significantly.

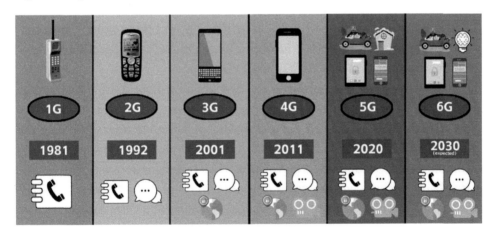

The widespread use of mobile communication enabled ubiquitous voice calls. But there were other points. The real communication revolution began with the rise of 3G Internet communication in the 1990s.

Internet technology was born in the 1950s. In February 1958, the US Department of Defense established a mysterious research department—ARPA

(Advanced Research Projects Agency) to gain a technological edge in the Cold War. Later, the Department of Defense decided to research a decentralized command system comprised of numerous nodes that could still communicate with each other even if several were destroyed. Thus, ARPA collaborated with several US universities to create the ARPANET, which was the precursor to the Internet.

During the 1960s and 1970s, ARPANET expanded and spawned the TCP/IP protocol, laying the foundation for the early Internet architecture. The 1970s saw the first boom of the integrated circuit era, initiated by transistors, marking the golden age of semiconductors. Advances in storage technology and leaps in chip capabilities led to an explosion in information technology. Not only did large computer systems mature, but personal computers were also born. All these nodes required network connections. In addition to the ARPANET-related technology line, Xerox's Palo Alto Research Center invented Ethernet, laying an important foundation for the development of data communication.

During this process, fiber optic communication, as an important technology, significantly impacted the later popularization of the Internet. In 1966, Chinese-American scientist Charles Kuen Kao published a paper predicting the emergence of fiber optic communication. Born in Jinshan County, Jiangsu Province (now Jinshan District, Shanghai), Kao was a physicist, educator, and expert in fiber optic communication and electrical engineering. He was a former president of the Chinese University of Hong Kong, a foreign member of the Chinese Academy of Sciences, and held British, US, and Hong Kong (China) citizenship. He was known as the "Father of Fiber Optics," "Father of Fiber Optic Communications," and the "Godfather of Broadband."

In 1970, Corning Inc. produced the first practical fiber optic cable with an attenuation value of 17 dB/km. Fiber optics enabled ultra-high transmission rates and bandwidth and were incredibly cheap. Although fiber optic communication is technically wired, it is still a form of wireless electromagnetic wave communication. It confines "free-roaming" electromagnetic waves within thin glass fibers, avoiding wireless communication's complex channels and interference, achieving ultra-high speed, advanced stability, and ultra-low latency. In essence, fiber optics are the true "foundation" of modern human communication networks.

Entering the 1990s, the Internet prototype had already formed, and the continuous expansion of its scale followed. At that time, developed countries in Europe and America had already entered the Internet era. The explosion of the Internet had a significant impact on mobile communication. A key feature of 3G was its support for data services. A series of network element reconstructions took place on the network side, transitioning from circuit switching to packet switching. Mobile communication began to serve multimedia data services. This transformation also ushered in the era of mobile Internet.

The mobile Internet fundamentally changed human lifestyles regarding clothing, food, housing, and transportation. It created countless new business models and built an online digital world. Image, audio, and video services, and online-to-offline businesses continually generated traffic. And traffic, in turn, stimulated further network expansion.

Following this came the 4G and 5G eras. The story of this phase is well-known to many. From 1G to 4G, from the user's perspective, 1G introduced mobile calling, 2G popularized it, 3G achieved mobile Internet, 4G provided faster Internet access, and 5G offered even faster Internet access, fulfilling almost all Internet needs and even beginning to venture into the IoT.

From the perspective of operators and mobile communication networks, from 1G to 5G, there has been a progression from low to high frequency and low to high speed. The system's capacity continuously increased, and its security and stability also improved while costs continually decreased. Ultimately, communication transformed from a privilege for a few to a benefit for everyone.

5.2 The Unprecedented Era of 6G Communication

Humanity's progress never halts, and the development and evolution of communication technology are no exception. Today, we stand once again at a turning point in communication technology, racing toward an unprecedented era of 6G.

5.2.1 The Era of 6G: What Does "6" Stand For?

For the average user, network speed is the key performance indicator (KPI) for each generation of networks.

In the 4G era, data transfer rates were tens of megabits per second. With 5G, data transfer rates are expected to increase by more than tenfold, reaching up to 1 Gbit/s (gigabit per second). When 6G arrives, network speeds will experience yet another leap. Technically, each generation's network speeds are set to surpass the previous one by an order of magnitude. Perhaps in the 6G era, network speeds might reach ten times or more than that of 5G.

So, what can be achieved with large bandwidth and high-speed rates? The famous movie *Ready Player One* is set in 2045. The protagonist, staying at home and wearing virtual reality equipment, can travel the world and join friends in cooperative battles, experiencing the ultimate immersive gaming experience. Technically, to achieve such effects, we would need the support of 6G.

Beyond fulfilling common needs like ultra-high-definition network TV and virtual reality games, holographic technology is a direction the industry is bullish on. The ITU has also included it in its white paper Technology Trends for 2030 and Beyond. Holographic technology refers to not only seeing images and hearing sounds but also achieving full sensory perception, including touch and taste. For example, we can be transported to a tropical rainforest thousands of miles away without leaving home through holographic imaging. We can see waterfalls, hear the sound of flowing water, and even feel the local temperature and humidity and smell the fragrance of the soil. In holographic remote meetings, we could touch each other; in holographic surgery, a remote surgeon could feel as if they were physically present.

However, holographic technology demands extremely high requirements for the number of sensors and data transfer rates. Richard Li, Chair of the ITU 2030 network technology research group and Huawei America's Chief Scientist for Future Networks, mentioned in a report that to achieve high-definition holographic projection of a person, the transmission rate must reach 4.62 Tbit/s—hundreds to thousands of times the speed of 5G, posing a challenge even for 6G.

The magic of 6G also lies in its extensive connections among people, objects, and devices. In fact, the industry proposed three major application scenarios for 5G: high-speed, large bandwidth, ultra-reliable, low-latency communication, and massive machine-type communications. The theoretical latency of 5G is expected to drop from 50 ms in 4G to 1 ms. Massive machine-type communications refer to connecting up to a million sensors per square kilometer.

Traditionally, from 1G to 4G, emphasis was placed solely on high-speed rates. 5G is the first to propose three goals. However, only "high-speed, large bandwidth" has achieved a unified international technical standard, and commercial 5G primarily satisfies this scenario. The other two, ultra-reliable low-latency and massive machine-type communications, are progressing slowly.

From the IoT perspective, the 5G era has only opened the door to IoT. The China Academy of Information and Communications Technology's Wireless Institute, in its 6G Concept and Vision White Paper released in March 2020, pointed out that 5G communication targets are concentrated within 10 km above the Earth's surface, unable to achieve seamless "space-air-sea-land" coverage envisioned for communication. As future demands grow and technology iterates, the road to truly achieving IoT is long, and this is precisely what the 6G era aims to address.

For example, to build a complete vehicular network, the problem of wide coverage must first be solved. Imagine an autonomous vehicle traveling on a smart road but suddenly losing signal in a remote area, leading to collisions.

Due to technical limitations, current 5G technology independently addresses high-speed, large bandwidth, ultra-reliable, low-latency communication, and massive machine-type communications. However, in the future, 6G aims to cover two or all three scenarios simultaneously. For instance, an entire city's intelligent traffic system not only needs to process real-time data from vehicle-to-vehicle, vehicle-to-person, and vehicle-to-infrastructure communications, requiring large bandwidth but also ensuring low latency to prevent accidents. Furthermore, to truly realize smart factories, 6G technology is necessary. 6G could reduce latency to millisecond or even microsecond levels, gradually replacing wired transmission between factory machines and achieving a higher level of wireless and flexible manufacturing.

5.2.2 Why Is Satellite Communication Necessary in the 6G Era?

From the perspective of communication technology development, the overarching vision of 6G is to extend and upgrade based on the 5G vision. A critical aspect of this is constructing an integrated air-space-sea-land network that transcends regional, aerial, and maritime boundaries, achieving truly global seamless coverage.

Although 5G has already brought significant changes to our lives today, many view it as a turning point from the human-centric Internet to the IoT, marking a key technological shift from consumer to industrial Internet. However, 5G has a limitation: it must cover broader areas. It relies on ground-based stations for communication, like 1G, 2G, 3G, 4G, and 5G. This means that unnecessary communication entities increase the cost per capita in areas with very low population density. Moreover, deploying 5G infrastructure in mountainous areas is particularly challenging. According to Statista, as of July 2023, the global Internet penetration rate was 64.5%, with mobile communication systems covering only 20% of the land area and approximately 6% of the Earth's surface. Approximately 2.8 billion people worldwide still lack Internet access. Furthermore, natural disasters such as Earthquakes, tsunamis, and forest fires can destroy communication infrastructure, leading to complete network blackouts.

Therefore, to progress toward the 6G era and build on 5G technology, other means of communication are essential to realize 6G. Among these, satellite communication will be an irreplaceable component in constructing 6G, or rather, a core technology. Compared to 5G, satellite communication can provide services to Internet users in remote and underdeveloped areas, as well as offer mobile communication services to special users such as aviation and navigation under extreme conditions, complementing ground networks to achieve true global coverage and offering uniform communication services to users worldwide. As illustrated below, the 6G era based on satellite communication will usher in an era of global communication without dead zones.

In terms of network speed, high-throughput satellite technology is becoming increasingly mature. Using high-frequency bands, multiple spot beams, and frequency reuse significantly enhances satellite communication capabilities. It reduces the cost per bandwidth unit, meeting the needs of high-speed information services and greatly expanding application scenarios.

Currently, Starlink offers download speeds exceeding 30 Mbps, even reaching up to 60 Mbps; upload speeds vary but generally maintain around 10 Mbps. This means users can smoothly watch ultra-high-definition videos and even enjoy online gaming.

Additionally, satellite communication has lower latency. Compared to traditional fiber-optic transmission, satellite communication speeds are very close to the theoretical value of the speed of light, nearly one-third faster than current mainstream fiber-optic solutions. It can achieve lower latency in the tens of milliseconds, which is significant in industries sensitive to delays, such as in global cross-border financial transactions. The professional market research firm TABB assessed that the "London-New York" route using Starlink satellites is 15 ms faster than ground-based fiber optics. This marginal communication latency advantage can bring considerable benefits to the financial industry.

In essence, without satellite involvement, there can be no true 6G. With satellites, coverage of vast celestial, aerial, and maritime areas would be possible, meaning forming a global communication mode would be unfeasible. With satellites, larger communication bandwidth and lower latency can be achieved. Only with satellite communication can humanity truly build an integrated air-space-sea-land network and stride into an unprecedented era of 6G.

5.3   5G NTN: The Gateway to the Future in the 6G Era

To transition from 5G to 6G, a natural development trend is integrating 5G with satellite communication.

On the one hand, 5G requires the complement of satellite communication. In regions like mountains, deserts, oceans, and skies, where ground network construction is difficult, complete network coverage is unattainable. Therefore, satellite communication becomes key in enabling seamless network connectivity and communication space extension for 5G in these areas. Compared to terrestrial mobile communication technology, the most significant advantage of satellite communication is its all-encompassing, terrain-agnostic, and distance-independent coverage. Additionally, satellite communication systems can provide continuous, uninterrupted network connection services, significantly enhancing 5G mobile communication technology applications in IoT devices and mobile users in airplanes, ships, trains, and cars.

On the other hand, as satellite communication connects isolated ground, sky, and sea networks into an integrated network, it also needs to leverage 5G's high transmission rate to enhance the user experience of LEO satellite systems.

It is now an industry consensus that 6G equals 5G plus satellite communication, which is also a necessary choice for the future of the communication industry.

5.3.1 How Can 5G Networks Integrate with Satellite Communication?

In fact, research on the integration of satellites and terrestrial mobile communications began in the 1990s.

The world's first regional satellite mobile communication system, the North American MSAT system, adopted analog ground mobile cellular network technology during construction, allowing for more seamless interaction between satellites and ground stations.

During its design process, Thuraya's satellite communication system adopted a Geostationary Mobile Radio interface similar to GSM/GPRS, allowing Thuraya to integrate better with terrestrial mobile communication systems while retaining GSM/GPRS standards. As a result, Thuraya could better adapt to user demand for mobile communication, providing more flexible and efficient services.

Well-known systems like Iridium and Global Star designed their air interfaces concerning GSM and IS-95. These satellite systems achieved higher interoperability with terrestrial mobile communication systems by adopting similar air interface standards. Users could use devices and protocols similar to terrestrial networks for smooth transitions to satellite communication systems.

By deploying ground auxiliary stations, the US SkyTerra and TerreStar systems enabled satellites and ground stations to operate in the same frequency band with nearly identical air interface signal formats. Terminal devices could seamlessly switch between satellites and ground stations, allowing users to connect to 4G wireless broadband networks in different environments without needing dual-mode terminals. This seamless switching design offered a more flexible and efficient communication experience.

In the 5G era, there are mainly two ways to achieve the integration of 5G and satellite communication. The first is to adopt a unified air interface protocol for both satellite communication networks and 5G networks from the design phase, with coordinated service management between space-based and ground-based networks. The terminal adopts a highly integrated design to seamlessly switch between LEO constellation networks and 5G communication networks, similar to the US SkyTerra and TerreStar systems. The second way is for satellite and terrestrial networks to maintain independent core network structures, with business data transmitted via satellite relay to ground gateway stations and then distributed by terrestrial communication networks.

Comparing the two methods, the former has greater advantages. Although adopting a unified air interface protocol is more complex and technically challenging than independent 5G networks and satellite communications, this method can enable satellite communication networks and 5G networks to achieve a higher degree of coordinated service through a unified protocol. Moreover, terminals with highly integrated designs can seamlessly switch between different networks, providing users with a smoother and more consistent communication experience.

Integrating 5G networks with satellite communication, known as 5G NTN, has already made some progress.

5.3.2 What Is 5G NTN?

What exactly is NTN? According to the standard definition given by the 3GPP, it refers to a network or network segment that uses airborne or spaceborne vehicles to carry relay nodes or base stations for transmission equipment. In simpler terms, it's a collective term for any network involving non-terrestrial flying objects, including satellite communication networks and high-altitude platform systems.

Satellite communication networks encompass spaceborne platforms like LEO, MEO, GEO, and GSO satellites. In contrast, high-altitude platform systems include airplanes, airships, hot air balloons, helicopters, drones, etc. Given that the current focus of 3GPP's work revolves around satellite communication networks, NTN, in a narrow sense, primarily refers to satellite communication.

5G NTN, then, is easy to understand. It refers to a new communication model in the 5th Generation Mobile Communication (5G) era, where ground cellular networks and satellite communication achieve high integration and collaborative service.

There are two common architectures for 5G NTN technology: transparent payload, also known as transparent forwarding, and regenerative payload, also referred to as base station in space. Transparent forwarding essentially uses the satellite as a signal relay link, with 5G base stations deployed behind the gateway stations as part of the ground network. The satellite does not process the signal transmitted from the base station but only connects mobile phones to gateway stations. In contrast, the base station in space is akin to deploying a 5G base station on the satellite. The ISL are like the Xn interface between ground base stations, and the feeder link between the satellite and the gateway station is effectively part of the backhaul network between the base station and the core network. The satellite is equipped with base station functions and can process data before transmitting it to the terminal.

Regardless of the architecture, our mobile phones will still interact with 5G base stations as they do now, except that some or all of these base stations will not be on nearby transmission towers but on communication satellites.

The benefits of 5G NTN are obvious. Unlike traditional 4G terrestrial networks, whose coverage is limited by terrain and natural disasters, 5G NTN offers mobile and fixed terrestrial operators a broader coverage range, enabling network coverage in oceans, deep mountains, disaster sites, and war zones. It's a significant breakthrough for remote communities, industrial sites, and cellular base stations, meeting more needs in more scenarios.

Currently, over half of the global population in certain regions cannot access the Internet. Including 5G NTN will provide robust support for remote communication, education, healthcare, disaster prevention, and reduction, significantly improving urbanization levels and economic conditions and promoting regional prosperity, thereby meeting the practical needs of governments and people in less developed areas. SpaceX believes that the "Starlink" project, once fully implemented, is expected to generate annual revenues of $30 billion globally. Moreover, the IoT is an inevitable trend, but existing communication methods are not up to the task. 5G NTN can ensure timely and effective data transmission, making IoT connectivity possible and enabling integrated management, control, and operation of all things. With 5G NTN, we need not worry about network connectivity wherever our devices are installed or used on Earth.

Furthermore, whether in LEO or MEO, satellite systems can provide high-speed, high-bandwidth services. With the rapidly increasing capacity of a new wave of satellite systems and maturing technology, the cost of accessing satellites has significantly decreased, supporting a broader promotional coverage.

5.3.3 The Development of 5G NTN

Currently, the integration of 5G and satellite communication has garnered extensive attention globally, including from standardization organizations like the ITU and the 3GPP. Especially with the introduction of 5G NTN, it has become a focal point of these research institutions.

Specifically, in 2016, the ITU proposed the need for "next-generation mobile networks to allow users to access services anytime, anywhere" and started researching the ITU-R M. [NGAT_SAT], a standard in satellite access technology. The ITU outlined four typical application scenarios for terrestrial-satellite integration, including relay broadband transmission, data backhaul and distribution, broadband mobile communication, and hybrid multimedia services. They also defined key characteristics to support these applications. Additionally, the ITU is actively advancing work on frequency usage spectrum sharing and electromagnetic compatibility analysis between satellites and 5G.

Starting with its Release 14 standards in 2017, 3GPP began demonstrating satellite communication's advantages to terrestrial mobile communication systems. In its technical report TR22.822, released at the end of 2017, the 3GPP SA1 working group defined three use cases for using satellite access in 5G: continuous service, ubiquitous service, and extended service. It also discussed the requirements for new and existing services.

Currently, 3GPP is primarily researching satellite communication in 5G through the "NTN" project, focusing on deployment scenarios and air interface design for satellite communication in 5G.

In June 2022, the 3GPP R17 standards were frozen, with satellite communication being a notable feature. In this standard version, two working groups focused on standardizing NTN: mobile broadband NTN and IoT use cases for NTN. The former uses the 5G NR framework for satellite communication, enabling fixed wireless access to backhaul from the ground to satellites and providing low-rate data and voice services directly to smartphones. The latter focuses on supporting low-complexity eMTC and NB-IoT LPWAN terminals for satellite access, expanding network coverage for various scenarios such as global asset tracking.

Following this, ABI Research predicted that satellite communication's annual compound growth rate would be 59% from 2024 to 2031. 5G NR-based NTN satellite services, expected to launch in 2026, will be a major driver for growth in NTN mobile connections. Another study on the satellite communication market by Juniper Research suggests that from 2024 to 2030, telecom operators could earn an additional $17 billion from 3GPP-based 5G satellite networks, with 5G NTN-based satellite communication connections reaching 110 million by 2030.

With the implementation plans for NTN in the 3GPP R17 set, various chipset, terminal, and satellite network operators have continuously released progress in satellite communication since the second half of 2022, leading to a surge in interest in the communication industry. According to ABI Research, in the past year, representatives from the communication industry like Apple,

Huawei, ZTE, Qualcomm, Motorola, MediaTek, and satellite operators like Globalstar, Inmarsat, and Iridium have intensified their strategic partnerships, promoting the integration of cellular and satellite communications. Many mobile communication operators are also joining, hoping to expand their network coverage through partnerships with satellite operators. Internationally, US operator T-Mobile collaborates with Musk's Starlink to provide satellite-to-mobile communication services. AST SpaceMobile and Lynk Global have established strategic alliances with several operators, including Vodafone, Rakuten Mobile, AT&T, Bell Canada, MTN, Orange, and Telefónica.

Shortly after the freezing of 3GPP R17, Ericsson, Thales, and Qualcomm, three overseas giants, announced the launch of a 5G space project, planning to develop a satellite network to provide high-speed connections for smartphone users regardless of location. The project tests 5G networks via near-Earth orbit satellites to help individuals in extreme geographical locations and remote areas access the Internet. The trio plans to launch satellites within four to five years at a potential cost of 8 billion euros, expecting speeds between 4G and 5G at several tens of Mbps.

This project relies on fewer but larger satellites for global coverage than Starlink, eventually requiring 600–800 satellites. Ericsson, Thales, and Qualcomm mentioned that the initial tests aim to validate various technology components needed for 5G NTN, including 5G smartphones, satellite payloads, and ground 5G network components. This work also seeks to confirm that 5G NTN can be supported within the elements of 5G smartphones, thus integrating future 5G smartphones with satellite phones.

Ericsson plans to validate 5G virtual wireless access network technology; Thales aims to validate a 5G satellite payload suitable for deployment on low-orbit satellites. Using its smartphones, Qualcomm will verify that future 5G smartphones can access 5G NTN. Ericsson, Thales, and Qualcomm will use ground-based equipment to simulate the propagation speed and delay between orbiting satellites and 5G smartphones.

For vendors in the 3GPP camp, as a 3GPP-supported technology, 5G NTN can leverage a large ecosystem of standardized products and components, offering technology suppliers opportunities for rapid expansion and compatibility between devices. According to an NTN report by ABI Research, by 2030, the global number of non-terrestrial network mobile connections is expected to reach $175 million, and the annual market size of global satellite services is projected to reach $124.6 billion. ABI Research views 3GPP's increased focus on satellite communication as having a profound impact on the entire satellite communication industry, with various satellite communication operators seeking market opportunities to integrate satellite communication with terrestrial cellular networks.

Apart from the rapid development of 5G NTN as defined by 3GPP, the integration of 5G networks with satellite communication has also progressed. In June 2017, Europe established the SaT5G (Satellite and Terrestrial Network for 5G) alliance, with members including BT, SES, Avanti, the University of Surrey, and other European enterprises and research institutions. The goal is to provide a cost-effective, plug-and-play satellite solution for 5G, offering continuous market growth opportunities for the satellite industry chain. At the European Networks and Communications Conference held in Ljubljana, Slovenia, in 2018, five SaT5G members demonstrated the integration of satellites with 3GPP network architecture, including VT iDirect and SES.

Additionally, to address the 5G system's goal of a 1000x capacity increase, the SANSA (Shared Access Terrestrial-Satellite Backhaul Network enabled by Smart Antennae) project, funded by the EU's H2020 program, aims to provide an excellent backhaul solution for future high-capacity wireless communication systems. The SANSA project proposes low-cost smart antenna beamforming technology, dynamic intelligent wireless resource management for terrestrial-satellite hybrid wireless networks, and database-assisted dynamic spectrum sharing technology and conducts in-depth research in these areas.

5.3.4 Technical Challenges to Be Addressed

It's noteworthy that, despite the progress in integrating 5G networks with satellite communication, numerous technical challenges still need to be overcome to truly realize this ambitious vision of integrating 5G networks with satellite communication essential for achieving a terrestrial-extraterrestrial integrated communication system for the 6G era.

5.3.4.1 Transmission Challenges

Looking at the development of 5G NTN, the first issue is the challenge of transmission—is it feasible to have satellite communication on the phone without enlarging the phone's antenna?

Direct satellite communication with phones means the phone can directly receive satellite signals. If the satellite's power is strong enough and the phone is sufficiently sensitive, it can pick up signals transmitted by satellites. The higher the satellite, the harder it is for the phone to receive the signal. The ability of a phone to achieve higher satellite communication rates depends greatly on its hardware, especially the type of antenna. Current phones have built-in antennae, while dedicated satellite phones have larger and more conspicuous antennae. According to antenna theory, an antenna's transmission and reception efficiency is high when its length is one-quarter of the radio signal's wavelength. More powerful than whip antennae are dish antennae, commonly seen as "dishes." A dish antenna can achieve good communication if accurately aligned. However, which phone would want to carry such a "dish"?

Therefore, under the current limitations of satellite bandwidth speeds, phones can only send texts or other small data and are incapable of high-speed Internet access. Huawei's Mate 60 series "direct satellite connection" and Apple's upcoming "direct satellite connection" are "conditional" direct connections. They connect to low-speed communication satellites and are conditional, low-frequency, narrow-band solutions for specific situations, supporting small data applications like text messages and short reports.

Then, in the 5G NTN stage, can the very tiny 5G NR (New Radio) in phones transmit signals to communication satellites? After all, the 5G NR in phones is designed for efficiency, working with antennae embedded in ultra-thin casings and connecting to base stations 5 km away, not 500 km away from a satellite.

Regarding this issue, Ericsson's Director of Technology Strategy, Per Synnergren, explained that, theoretically, current phones should be compatible with satellite base stations. However, Synnergren also added that achieving the best connection might require some adjustments to existing phone antenna designs. Battery consumption would be slightly higher than most terrestrial 5G connections. Notably, considering the physical scale involved, such a connection seems feasible. Many practical factors contribute to the viability of this work. For instance, 5G NR is already designed to handle very challenging conditions, including poor signal strength and multipath in large, crowded terrestrial base stations. Moreover, NTN could have its own spectrum segment, free from interference from terrestrial 5G networks or external signal sources.

5.3.4.2 Mobility Challenges

In addition to receiving signals from satellites, another challenge is mobility. Compared to the independent operation of terrestrial networks and satellite communication, mobility management in an integrated terrestrial-satellite network presents more severe challenges.

We know that satellites are always moving in the sky. Therefore, if a phone needs to use beamforming to concentrate its transmission energy on a satellite, it needs to know the approximate location of the relevant satellite at any given time. In other words, the phone requires an ephemeris or satellite trajectory chart to calculate positions and paths.

Next is the Doppler effect. During the connection process, as the satellite moves along its orbit at about 8 km/s while emitting signals, this movement will cause fluctuations in the frequency of signals between the phone and the satellite, similar to how the pitch of a siren changes as a fire truck speeds by. But today's 5G NR is designed to lock onto a frequency and stay there, not to track a moving frequency. So, in the future, this aspect may need some redesign.

Moreover, in terrestrial networks, we usually switch from one base station to another only when crossing regional boundaries or when signal quality severely degrades due to some obstacles. Since phone towers can be several kilometers apart, signal handovers don't occur often, even in moving vehicles. However, for 5G NTN, the cells defined by those massive antenna arrays on satellites are relatively small and continuously sweep across the Earth's surface. A stationary phone might stay in a particular cell for only six to seven seconds. This means that the phone, base station, and core network must be handed over to a new base station frequently every few seconds.

Besides the movement of satellites, there is also the issue of propagation delay. The goal for terrestrial 5G networks is a round-trip delay of 1 ms between sending information to a base station and receiving a response. The number of critical timers in the radio control parts at both ends depends on this number. However, radio waves take 30–50 ms to travel from the ground to a near-Earth orbit satellite. Such a delay would at least require rewriting some control firmware, which could make services like network gaming completely unviable. This issue is also one of the most urgent to address for the minimum latency services that 5G provides to IoT devices.

5.3.4.3 Regulatory Challenges

While technological challenges seem manageable, business and regulatory issues are comparatively more complex.

From a business perspective, one prominent question is whether enough users are willing to pay for global 5G NTN coverage. This forms a crucial commercial consideration. Beyond specific users like high-income global travelers, military personnel, and adventurers, would the pricing of 5G NTN services be attractive to ordinary users? And for those who genuinely need 5G NTN global coverage but may not receive good service, the affordability of such services is a practical and significant issue. After all, the cost of satellite launches and maintenance needs to be shared among users.

Another commercial issue is whether potential users capable of paying for high-quality services would be satisfied with early 3G-level services instead of pursuing more advanced data services. This not only requires in-depth market surveys and research but also an accurate understanding of user needs. Like technical issues, this problem also demands significant time and resource investment.

The commercial model challenges for 5G NTN extend further. The economic viability of satellite services requires building a large-scale user base globally, and whether this is achievable remains to be seen based on market feedback. Additionally, differentiated pricing strategies, service packaging, and promotional activities for various user groups are key strategies that need careful consideration and formulation in the commercial realm.

From a regulatory standpoint, numerous challenges also exist. One of the core regulatory issues is determining who has the authority to manage specific base stations, especially when these base stations cross national borders. Addressing this issue involves establishing effective international regulatory bodies or collaborative frameworks.

First, which entity should have jurisdiction over base stations? The need for an international management body to coordinate and oversee the operation and maintenance of transnational base stations is an urgent topic. Current regulatory systems focus mainly on the national level, but the emergence of transnational base stations makes international regulation increasingly urgent. Establishing an international management body requires consensus in international cooperation and governance to balance and coordinate interests among nations.

Second, the distribution of control and regulatory responsibilities for base stations crossing national borders becomes a complex issue. Different regulations, standards, and regulatory systems may exist between countries, making it challenging to achieve clear base station control in such situations. This might involve formulating transnational agreements to ensure legal and orderly operation and management of base stations across borders. National control over spectrum allocation also poses an unavoidable regulatory challenge. With satellites potentially needing to switch frequencies during transnational transmission, managing spectrum at national borders becomes extremely difficult. How to coordinate and resolve international issues regarding spectrum allocation is a direction requiring joint efforts from governments and international organizations.

Furthermore, more and more countries are controlling data communication within their borders, including Internet access, VPN usage, and data privacy. This poses additional regulatory challenges for 5G NTN services. Addressing this issue requires international cooperation to establish common data privacy standards and Internet access policies, ensuring the security and legal transmission of user data.

In summary, 5G NTN presents an appealing vision of expanding 5G service coverage and bringing new business opportunities to terrestrial communication service providers. Research by technology developers like Ericsson, communication and satellite equipment manufacturers, and service providers like T-Mobile and SpaceX, may soon inform us about the technical feasibility and associated costs. However, this might just be opening a Pandora's box filled with complex business and regulatory issues, which may take longer to resolve. The balance between being a pioneer and assuming unquantifiable risks in this opportunity is very delicate.

Looking ahead, the future remains certain and foreseeable. With the acceleration of satellite network formation, satellite communication will be integrated into the entire 5G ecosystem. Communication technology will further advance beyond 5G toward 6G, changing our communication methods with more powerful connections, broader coverage, and more innovative applications, leading humanity toward an integrated 6G era of space, air, land, and sea. By then, satellite communication will not just be an independent means of communication but will be deeply integrated with terrestrial networks, building a more complete and efficient communication ecosystem.

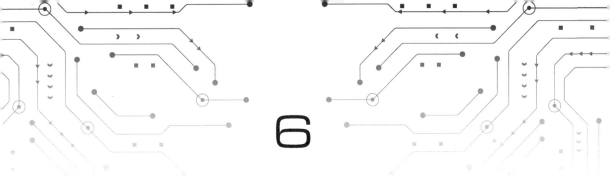

6

TOWARD THE 6G ERA

6.1   Satellite IoT: The True Era of Connecting Everything

With the acceleration of satellite communication construction, a new era of the Satellite IoT is dawning.

In fact, IoT is not a new concept; it has existed since the era of the Internet. Today's IoT devices primarily rely on cellular networks. Although cellular network infrastructure is more widely available now than a decade ago, it still faces the limitations of terrestrial networks. In many parts of the Earth, such as remote mountainous areas, the depths of oceans, and polar regions, traditional IoT technologies struggle to provide coverage due to remote locations and harsh environments, severely limiting information access and communication capabilities in these areas.

Meanwhile, the demand for global coverage, real-time communication, and high-quality transmission is increasing. To realize the vision of IoT, we need to break through the limitations of terrestrial networks. This is where Satellite IoT comes into play.

6.1.1 Everything Can Be Connected

The concept of the IoT was introduced by Kevin Ashton in 1999. Literally, the IoT refers to connecting objects through networks. Essentially, these objects can be anything in the world, encapsulating Ashton's vision that "everything can be connected."

The earliest objects to be networked likely came from the military domain, such as drones and unmanned vehicles. These objects require real-time information acquisition and transmission to accomplish various tasks, like reconnaissance and exploration.

Of course, objects worth networking extend far beyond these. From an informatization perspective, many objects that people take for granted also need to be networked. Not just inanimate "things," but living organisms, including plants, animals, and even humans themselves, can be considered "objects" for networking purposes. However, the IoT for living organisms, including humans, differs from the Internet networking era.

For example, buildings and houses are common "objects." While people might pull fiber optic and ethernet cables into houses for networking, the structure of the house itself rarely becomes the object of networking. Under the IoT paradigm, people consider embedding sensors within the structure of buildings to measure pressure, tension, shear force, deformation, temperature, humidity, etc., integrating them into steel structures or concrete. Thus, a house truly "comes to life" as if it had its own life force.

This enables building safety and public utility units to monitor the safety of houses through such sensors, which is crucial for homes in geologically unstable areas. If unsafe modifications are made to a house, they can be quickly detected and addressed.

Beyond houses, cars, construction machinery, wildlife conservation and monitoring facilities, remote electrical facilities, oil and gas pipelines, maritime lighthouses, and buoys can all become networked "objects." For instance, vehicles and construction machinery can monitor their operational status and health through built-in sensors, enhancing maintenance efficiency and safety. Wildlife conservation and monitoring facilities can track the movements and ecological information of wildlife in real-time, providing data support for conservation efforts.

In the industrial sector, IoT applications are also significant. Through real-time monitoring, companies can obtain more accurate production data, analyze it to identify potential issues, and make timely adjustments. For example, a manufacturing company deploying sensors along production lines to monitor equipment status and product quality can leverage IoT platforms for data analysis and adjust production parameters to ultimately enhance product quality and efficiency.

Predictive maintenance represents another important application of IoT in the industry. By remotely monitoring and diagnosing equipment status, companies can identify potential failures in advance, undertake predictive maintenance measures, and avoid production downtime and increased maintenance costs.

IoT can also optimize supply chains by integrating sensors and IoT devices into logistics, enabling real-time tracking of material transportation and storage conditions, thus improving supply chain visibility and collaboration. For instance, a manufacturing company installing IoT devices on cargo vehicles to monitor the location and status of goods in transit can effectively reduce losses and delays, enhancing the efficiency of the supply chain.

Turning to living "objects," such as the human body, can the human body be considered an "object"? Indeed, it can. We know that many parameters describe a person's health, most of which are unlikely to be determined by our own senses. Even professional healthcare workers need the aid of various instruments to measure health-related parameters. However, with the advancement of sensor technology, people can now measure many vital signs using wearable devices, including heart rate, body temperature, and blood oxygen saturation.

Wearable devices, through built-in sensors, can measure multiple vital signs, such as heart rate, body temperature, and blood oxygen saturation. This data is transmitted to cloud servers, allowing medical personnel or individual users to access this information anytime, anywhere. For the elderly, timely monitoring of health conditions is particularly crucial due to gradually declining physiological functions. Elderly individuals often face multiple health issues, such as chronic diseases, cardiovascular diseases, and diabetes. IoT technology enables real-time monitoring and management of these conditions. For example, monitoring heart rates for cardiac patients and blood sugar levels for diabetics can detect abnormalities early, allowing for timely medical intervention. This underscores the significance of networking humans as "objects."

6.1.2 Breaking the Coverage Bottleneck of IoT

Understanding the "things" in IoT, we now turn to the challenge of networking them. In cities, networking is easy thanks to fiber optics and mobile communication towers. Beyond cities, national communication backbones connect nearly all cities and towns. However, in broader areas like deserts, mountains, forests, lakes, tidal flats, and oceans, such ideal communication conditions do not exist, making it difficult to establish IoT.

We all know that our planet, Earth, is more of a "water planet" than a "land planet." About 71% of Earth's surface is covered by oceans, with only 29% being land. Even on land, after years of development, communication networks only cover about 20% of the land area. And in the oceans, coverage is even less, at only about 5%. In other words, less than 10% of Earth's surface has network signal coverage.

This lack of coverage brings various problems. A typical example is flying from China to the US. After leaving China's air traffic control area, flights pass over the wilderness of Siberia, then over the Arctic Ocean, to Alaska, US, and then into the desolate areas of northern Canada, only getting network service near the US-Canada border. Therefore, one of the biggest frustrations for people on long-haul flights is the inability to access the Internet.

Oceans are areas where it's impossible to build fiber optic networks and towers. Especially considering that oceans cover 71% of Earth's surface, the presence or absence of communication networks determines humanity's ability to explore, utilize, and protect Earth's resources.

Moreover, groups of people need communication networks in the ocean. One group is seafarers, as most global bulk goods are transported by sea. Sea transport accounts for over two-thirds of international trade volume, with over 90% of China's imports and exports being conducted by sea. Without sea transport, many places would struggle to access sufficient oil, food, coal, and daily necessities. To serve sea transport, a large number of seafarers work on ships. However, they often live in isolation, finding it difficult to contact family when the ship is not docked. Although modern ships have maritime satellite phones, such satellite communication services are very expensive, and seafarers only make emergency calls.

Another group that lives at sea for extended periods includes workers on various oil drilling platforms. As the locations of these platforms are relatively fixed, VSAT satellite terminals are installed on the platforms for communication with GEO satellites. The Internet conditions on oil drilling platforms are much better than on merchant ships, but the use of traditional GEO satellites is also costly, allowing oil workers only narrowband network services, making high-definition video calls with family a luxury.

Besides the familiar oil drilling platforms, merchant ships, fishing boats, cruise ships, and passenger ships, there are also deep-sea aquaculture bases. On these large deep-sea aquaculture processing ships, not only do many crew members need to contact family and enjoy entertainment through the network, but they also need to connect to e-commerce networks to purchase supplies and sell fish catches.

So, the question arises: if the coverage is so low, how can we "connect everything"? How do we implement IoT applications in places without terrestrial communication system signals? This is where the Satellite IoT comes in—leveraging the broad coverage capability of satellites to compensate for the shortcomings of terrestrial communication systems, providing wider connectivity. After all, satellites are in space, and as the saying goes, "standing high allows one to see far." Their coverage capability is incomparable to terrestrial base stations or fiber optic cables. Satellite communication can truly shine, especially in areas like deep mountains, forests, deserts, grasslands, and Gobi, where traditional terrestrial communication is hard to reach. This enables the realization of a truly global IoT in every corner of the Earth.

6.1.3   A Complete IoT

In the previous chapter, we mentioned that the finalized 3GPP R17 standard introduced the concept of NTN, positioning satellite communication networks as an essential supplement to terrestrial networks. Besides providing 5G NR NTN standards for satellite communication, R17 also offered standard specifications for integrating IoT technologies like NB-IoT and eMTC with NTN.

Both ABI Research and Juniper Research predict that by 2030, the number of 5G NTN connections will exceed 100 million, and a majority of these connections will likely be IoT-related. Compared to NR-NTN, supporting broadband data and voice for mobile phones, IoT-NTN, supporting the IoT, seems to be progressing faster toward commercial implementation, from devices to testing to application scenarios.

During MWC in late February 2023, MediaTek showcased its MT6825 IoT-NTN chip, an independent SoC and RF solution designed as an attachable solution for existing smartphone designs. Utilizing common L&S band satellite frequencies, it provides narrowband satellite connectivity, connects to GSO satellites, and is easily adaptable for use with 3GPP NTN standard satellite networks. Additionally, the MT6825 allows devices to automatically receive information from satellites, offering a seamless satellite communication experience.

In June, Qualcomm introduced two modem chips designed specifically for IoT satellite communication, the 212S, and 9205S, supporting 3GPP R17 IoT-NTN. The 212S is designed for fixed IoT terminals in remote locations and fixed applications, such as telemetry data collection for infrastructure devices like water, gas, and electricity meters, as well as utility grid monitoring, initial fire investigation reports, terrestrial and maritime mining devices, and environmental management. The 9205S, on the other hand, is designed for mobile IoT terminals and caters to industrial applications that require always-online, terrestrial and satellite hybrid connectivity for goods and assets in motion, such as transoceanic shipping container tracking, agricultural equipment and livestock tracking, and global fleet and freight tracking for supply chain management. Qualcomm previously launched Qualcomm Aware, a platform for IoT asset tracking, and the 212S and 9205S IoT-NTN chips are also integrated with this platform, providing real-time asset tracking and terminal management functions in remote areas.

UNISOC also launched the first satellite communication SoC chip, V8821, compliant with the 3GPP R17 IoT-NTN standard. Utilizing L-band maritime satellites and S-band Tiantong satellites, it offers data transmission, text messaging, calls, and location-sharing functions. It can also be expanded to support access to other high-orbit satellite systems. It broadly applies to communication needs in areas like oceans, urban edges, and remote mountain regions where cellular networks struggle to reach.

Regarding application scenarios, satellite IoT has already found ground in areas not covered by networks, tracking, controlling, and managing devices and living organisms, including agriculture, electricity, construction sites, ocean-going ships, offshore oil wells, and global supply chains for multinational corporations. For consumers, satellite IoT can provide emergency terminals for outdoor enthusiasts, offering timely location and vital sign feedback. Additionally, satellite IoT can be used in research, environmental protection, and animal tracking.

For instance, Orbcomm's maritime container tracking management requires users to purchase tracking and IoT terminal devices for continuous cargo and container tracking, collect temperature data and other information, and use cellular, LoRa, and satellite IoT networks for data transmission. Orbcomm also provides platform services for data analysis applications. With real-time container tracking, Orbcomm helps optimize container capacity utilization, reducing empty load and capacity loss. For special transportation scenarios like cold-chain containers, Orbcomm enables real-time, all-weather monitoring of cargo conditions like temperature, reducing transportation loss.

Another example is the forest fire monitoring project by Guodian Gaoke, deploying integrated terminals with sensors in forest areas, achieving near real-time communication of sensor information via low-orbit satellite IoT. Sensor data includes temperature, humidity, smoke detection, vegetation, brightness, atmospheric pressure, and meteorology, and also supports ground wireless data transmission. This enables early detection of potential fire sources in sensitive forest fire prevention areas without visible flames.

Currently, Starlink has also launched satellite IoT services. Swarm Technologies, fully acquired by SpaceX in September 2021 and becoming a wholly-owned subsidiary under the Starlink division, now offers narrowband satellite connections via near-Earth micro and nanosatellites at a reduced cost of $5 per month. Named "Space Bees," the Swarm constellation consists of 150 satellites operating in near-Earth orbits of 450–550 km and can be widely applied in agriculture, maritime, energy, environment, and transportation sectors, providing reliable connections to remote areas.

Looking to the future, satellite IoT will bring even more application scenarios. For example, in logistics delivery, convenient and developed logistics networks are not only an essential part of modern production and life but also a vital support for new rural construction in remote areas. With dual-mode terminals integrating terrestrial and satellite communications, seamless global transmission and remote control of vehicles and cargo can be achieved, aiding in the construction of the last mile of logistics. In the field of power transmission, transmission lines are distributed in towns, villages, mountains, and remote areas. Through terrestrial networks and satellite IoT, dual-mode IoT terminals enable seamless, continuous monitoring of power line parameters.

While satellites still face numerous technological complexities and challenges, such as path loss, Doppler shift, security, interoperability, cost, etc., the proliferation of low-cost, low-power global satellite connections will accelerate the growth of connected sensors worldwide. Satellite-based IoT will efficiently connect many devices in every network unit, ultimately realizing a highly intelligent ecosystem.

6.2 Autonomous Driving: A Breakthrough Application in the 6G Era

The evolution of transportation has undergone several phases to date. From the earliest form of mobility—human feet—to domesticated animals like horses and donkeys, followed by carriages and carts. The advent of the steam engine brought automobiles and trains, radically changing transportation. Modern civilization's rapid progression made traveling through air and water commonplace. The advent of the 6G era promises yet another transformative shift in transportation, with autonomous driving playing a pivotal role.

Autonomous driving, a staple of science fiction novels and movies, represents a top-tier technological engineering challenge closely linked to everyday life. The terms "difficult" and "disruptive" are inextricably linked to the concept of autonomous driving. So, how far has the development of autonomous driving cars come today? And why are they considered a key transformative element in the 6G era?

6.2.1 The Development of Autonomous Driving

Rewind to August 1925: the world witnessed the debut of the first-ever autonomous car, the "American Wonder." No one sat in the driver's seat, and components like the steering wheel, clutch, and brakes seemed to operate independently. Engineer Francis P. Houdina controlled the preceding vehicle in the car behind by radio waves. Though more of a grand-scale remote control experiment, this journey from Broadway to Fifth Avenue in New York is not widely recognized in the industry due to its rudimentary understanding of automated vehicle mechanics.

At the 1939 New York World's Fair, the first concept car of autonomous driving, "Futurama," was unveiled by General Motors. It was an electric car guided by a radio-controlled electromagnetic field. GM's design involved magnetic spikes embedded in roads, creating electromagnetic fields to guide electric vehicles. Futurama's impressive showcase at the fair significantly popularized the concept of autonomous driving worldwide.

A year later, designer Norman Bel Geddes stated in his book *Magic Motorways* that humans should be removed from driving. He envisioned US highways equipped with train-like tracks providing an automatic driving system. Cars would follow set trajectories and programs on highways and switch back to human control when exiting them. He predicted this would be realized by 1960.

However, reality proved to be starkly different from these lofty ideals. It wasn't until the 1950s, when researchers began experimenting with these concepts that the true challenges became apparent. Nevertheless, the quest for autonomous driving technology persisted globally.

In 1966, intelligent navigation first emerged at Stanford Research Institute. Shakey, a robot with a wheeled structure, was developed by the SRI Artificial Intelligence Center. Equipped with sensors and software systems, Shakey pioneered autonomous navigation functionalities.

By 1977, Japan's Tsukuba Mechanical Engineering Laboratory enhanced General Motors' radio signal control approach, innovatively designing a camera system to process road images. This led to the birth of an autonomous passenger car that could follow white road markers at 30 km per hour. Though lateral control still relied on rail assistance, this car was significant and is considered the first modern autonomous passenger vehicle.

In 1989, Carnegie Mellon University in the US used neural networks to guide autonomous vehicles. Though their retrofitted military ambulance in Pittsburgh had a server as big as a refrigerator and a computational capacity only a tenth of an Apple Watch, the principle closely resembled today's autonomous vehicle control strategies.

Parallel to global developments, China also began research on intelligent mobile devices in the 1980s, initially focused on military applications. In 1980, the Remote Control Driving Anti-nuclear Reconnaissance Vehicle project was launched nationally, involving the Harbin Institute of Technology, Shenyang Institute of Automation, and the National University of Defense Technology.

Persistent efforts by researchers led to the National University of Defense Technology developing China's first prototype autonomous vehicle in 1987. Though it appeared similar to other cars, it possessed basic autonomous driving functions, marking a breakthrough in Chinese academia's exploration of this technology. In 1988, as part of the national 863 Program, Tsinghua University began developing the THMR series of autonomous vehicles, with the THMR-V capable of automatic lane following in structured environments.

In 2004, the US Defense Advanced Research Projects Agency (DARPA) initiated the DARPA Autonomous Vehicle Challenge to inspire top talents to join the development of autonomous vehicles—the world's first long-distance autonomous car race. Participants showcased innovative solutions, propelling the field forward. The challenge continued until 2007, fostering a generation of talents in autonomous driving technology.

Google began its foray into autonomous driving in 2009, recruiting many key participants from the DARPA challenge, including Stanford talents. This led to a continuation of the tradition of relying on technologies other than camera vision in Google's autonomous cars.

In 2012, the release of the KITTI data set marked a further maturation of autonomous driving technology, driving research in visual deep learning for autonomous driving. Covering GPS-RTK inertial navigation systems, stereo cameras, and LIDAR sensor data, the KITTI data set facilitated precise measurements and performance evaluations of visual algorithms. The data set's launch brought deep neural networks back into the autonomous driving scene, with computer vision and machine learning rapidly exploring the boundaries of this technology.

The realization of autonomous driving cars has been a lengthy journey, spanning nearly a century and encompassing multiple cutting-edge technologies, including AI. It's the cumulative effort of these advancements that has shaped the state of autonomous driving today.

6.2.2 The Need for 6G Communication in Autonomous Driving

The SAE's six levels of autonomous driving are no automation (L0), driver assistance (L1), partial automation (L2), conditional automation (L3), high automation (L4), and full automation (L5).

L0, "no Automation," represents the stage where the driver has absolute control.

L1, "driver assistance," allows the system partial control at a given time, either steering or throttle/brake. In emergencies, the driver needs to be ready to take over immediately and must monitor the environment.

L2, "partial automation," transfers control from "partial" to "full," meaning the driver can hand over both lateral and longitudinal control to the system under usual driving conditions. However, the driver still needs to monitor the environment.

L3, "conditional automation," signifies the system handles most driving operations, with the driver responding appropriately in emergencies. The system, not the driver, monitors the environment.

L4, "high automation," indicates the system performs all driving tasks even without driver response but only supports certain driving modes and not all scenarios.

L5, "full automation," differs from L0–L4 by supporting all driving modes. In this phase, drivers may no longer be allowed to control the vehicle.

Technological advancements indicate that current smart driving technologies globally are mostly at L2 to L3 levels. Notably, from L3, autonomous driving means the vehicle completely handles all issues during driving, including acceleration, overtaking, and even obstacle avoidance, shifting accident liability from humans to vehicles. L3 is pivotal in autonomous driving, bridging lower-level driving assistance and higher-level full automation.

The L2 level focuses on the human driver, with the autonomous system only assisting. L2 corresponds to ADAS (Advanced Driver-Assistance Systems) technologies, including ACC (Adaptive Cruise Control), AEB (Automatic Emergency Braking), and LDWS (Lane Departure Warning System), requiring the driver to remain in control.

L3, however, achieves "no human intervention," with the autonomous system handling most driving decisions and actions. The vehicle operates under specific conditions, but in emergencies, the driver still makes decisions. Functions include highway piloting (HWP, 0–130 km/h), traffic jam piloting (TJP, 0–60 km/h), automatic parking, high-precision maps, and precise positioning.

In October 2015, Tesla introduced its Autopilot driving assistance system, the first commercial driver-assistance technology. Tesla's production cars are equipped with Autopilot 1.0, 2.0, or 2.5 hardware systems, with autonomous driving capabilities upgradable from Level 1 to Level 4+ via over-the-air updates, preparing for further full autonomy.

By 2016, autonomous driving had become a focal point for many companies. Numerous enterprises, including traditional automakers like Dongfeng, Geely, BAIC, and SAIC, Internet giants like Baidu, Tencent, Alibaba, and ride-hailing services like Didi, entered the autonomous driving race. All actively participated in the development of autonomous driving technology.

In 2018, the new Audi A8 made its global debut, equipped with an L3 autonomous driving system. Dubbed "Audi AI Traffic Jam Pilot," the system activates with an "Audi AI" button, allowing full system control at speeds below 60 km/h. With the L3 system, the car autonomously performs driving tasks like acceleration, braking, and steering. Drivers can leave the steering wheel as the system manages specific driving scenarios.

However, whether Tesla or Audi, current autonomous driving remains at L3, struggling to break through to higher levels. Two key challenges are AI limitations and current network constraints. For instance, autonomous vehicles may not correctly judge when to overtake a car ahead. Incidents like Tesla's 2016 accidents in China and the US occurred due to the 4G network's inability to process sensor data at high speeds.

5G networks offer higher transmission rates and lower latency, improving the efficiency of data collection and feedback for autonomous driving. However, current cellular networks are only sometimes stable, with potential disconnections in tunnels or covered areas. This poses safety risks for autonomous vehicles, especially in remote areas with sparse cellular coverage.

Moreover, cars are increasingly integrated into daily life, serving as spaces where people spend significant time. Morgan Stanley estimates over 600 billion hours annually are spent in cars, a figure that could rise to trillions with further automation. To enhance user experience, autonomous vehicles require extensive data for software updates, navigation, route planning, and other functions. Both densely populated and remote areas need broadband networks, a demand 5G needs to meet.

6G networks, based on satellite communication, are essential for autonomous driving. They offer larger bandwidth and higher data transmission rates for real-time analysis and decision-making. The low-latency transmission of 6G is crucial for the instantaneity required by autonomous driving systems.

Additionally, the wide coverage of 6G, an advantage over traditional cellular networks, ensures global connectivity, vital for autonomous driving in remote areas. Satellite communication enables the transmission of massive data collected by sensors like HD cameras, LiDAR, and infrared sensors to the central system for immediate analysis.

Despite challenges like path loss, Doppler shift, security, interoperability, and cost, the proliferation of low-cost, low-power satellite-based global connections will accelerate the growth of connected sensors worldwide. Satellite-based IoT will efficiently connect numerous devices in each network unit, realizing a highly intelligent ecosystem.

6.2.3 Redefining Smart Traffic

Beyond smart driving, another key future vision of smart traffic is the realization of the "Internet of Vehicles" (IoV). The innovation of IoV lies in combining AI with the IoT to provide high-value services to drivers.

The electronic control modules and sensors throughout the vehicle facilitate "Vehicle-to-Vehicle" and "Vehicle-to-Infrastructure" connections. Supported by AI technology, vehicles can proactively suggest route changes, avoid road hazards, and request assistance in accidents. Soon, cars will accurately know their relative positions to other vehicles, identify potential dangers, and take preemptive actions to prevent accidents, paving the way for a more convenient and efficient shared transportation system.

However, this all relies on "connectivity," specifically 6G era connectivity. To achieve efficient vehicle interconnectivity and intelligent connections with road infrastructure, a robust and reliable network is essential—a network with absolute high speed, low latency, and wide coverage, all achievable only with 6G based on satellite communication.

Specifically, vehicles need to share large amounts of data in real-time, including location, speed, and direction. This data must be rapidly transmitted among highway vehicles to support real-time decision-making and coordinated actions. The 6G network based on satellite communication provides greater bandwidth and higher data transmission rates, meeting the urgent need for high-speed data transmission in vehicle interconnectivity. For example, through a 6G network, vehicles on highways can share traffic conditions in real-time, enabling intelligent fleet coordination and improving road efficiency. Additionally, near intersections, vehicles can share their intended movements via 6G, enabling efficient and safe intersection crossing.

Moreover, vehicles may travel in cities, suburbs, rural, or even remote areas, requiring global network coverage. Traditional cellular networks may have insufficient coverage in remote areas, while 6G networks based on satellite communication achieve global coverage through satellite constellations. For instance, in remote mountains or deserts, the 6G network ensures efficient communication between vehicles and infrastructure, enabling global vehicle interconnectivity.

Historically, transportation has evolved through several stages. From the earliest mode of foot travel, humans then utilized domesticated animals like horses and donkeys, as well as horse-drawn and ox-drawn carts. Later, with the advent of the steam engine, automobiles and trains replaced these primitive modes of transportation. To reach desired destinations, humans devised various solutions. For over a century, fuel-driven cars, operated by humans, replaced previous transportation means but brought endless problems.

For example, the US has 212 million licensed drivers and 252 million vehicles, traveling 5 trillion km annually and consuming over 700 billion liters of fuel. Cars and trucks account for 20% of the US greenhouse gas emissions. Despite this, most people still consider owning a car a necessity in modern society. However, in reality, cars are idle 95% of the time.

This means cars are only used 5% of the time, requiring parking spaces for the remaining 95%. Consequently, people's homes, workplaces, shopping centers, hospitals, stadiums, and streets all need to accommodate parked cars.

Furthermore, these vehicles have low energy efficiency. Less than 30% of the fuel in a car is used for propulsion, with a small portion powering lights, radios, and air conditioning, while the rest is wasted as heat and noise. The average car weighs about 1,400 kg, and the human body about 70 kg, meaning only about 5% of the energy used to drive a car is for transporting passengers, just 1.5% of the total fuel energy.

The low-efficiency results from cars being designed with performance far exceeding most daily needs, termed "occasional use requirements" by Waymo CEO John Krafcik. In the US, 85% of travel is by car, with an average occupancy of 1.7 people per vehicle, dropping to 1.1 during commutes. Urban average speeds are as low as 20 km/h.

Therefore, Morgan Stanley financial analyst Adam Jonas described cars as "the world's least utilized asset" and the automotive business as "the business that should stop on the planet." This is why Pulitzer Prize-winning journalist Edward Humes said, "Almost every aspect of car allocation and use is insane."

The future, however, offers a healthier transportation solution with the advent of autonomous driving and IoV based on 6G communication. Imagine a near future where most people no longer need to own or drive cars but rely on safe and convenient autonomous vehicle services. Transportation service companies will provide comprehensive mobility services, eliminating the hassles of parking, cleaning, maintenance, and charging. The troubles associated with car ownership will disappear. People won't need to buy, mortgage, or insure cars, nor spend time driving, parking, or refueling. Traffic will no longer be a headache. With a simple phone tap, a car can be summoned. The vehicles without steering wheels, pedals, or brakes will mostly be comfortable two-seater electric cars. All this will significantly reduce future transportation costs and be more environmentally friendly.

As urban planning changes with transportation systems, a smart city with autonomous driving won't need as many parking spaces as today's cities. Models by the Organization for Economic Cooperation and Development suggest a city like Lisbon would need only as much land as 210 football fields for parking. Many existing garages could be converted to retail facilities, and the need for street parking will drastically decline. On the other hand, cities will need to provide large drop-off areas for robotic taxis.

More importantly, society will soon welcome a new transportation revolution—the era of flying cars. Autonomous driving is a core technology for flying cars, which, aside from AI-based intelligent driving technology, also relies on satellite technology for communication and precise navigation.

People's expectations for fast, reliable, convenient, and personalized travel options rapidly evolve with communication technology. Under this trend, people's travel habits and behaviors are undergoing significant changes. In the foreseeable future, the era of flying cars will unfold new possibilities beyond our current imagination.

6.3 Satellite Communication: Sparking Military Transformation

In the vast changes by satellite communication, the military is one of the most affected domains. Historically, the evolution of communication technology has continually shaped military operational tactics, where the flow of information often determines the speed, momentum, and precision of actions. At the same time, distance, weather, and terrain try their utmost to impose limitations. However, 6G based on satellite communication can maximize the removal of these constraints, providing truly resilient connectivity for the battlefield.

6.3.1 The Use of Starlink in the Russia-Ukraine Conflict

In the Russia-Ukraine conflict, LEO satellite communication systems, exemplified by Starlink, emerged prominently in warfare.

In February 2022, Russian military forces entered Ukraine to conduct special military operations. Due to network disruptions in Ukraine caused by the conflict, the Ukrainian Deputy Prime Minister appealed to Elon Musk for Starlink assistance to maintain connectivity. Musk responded to the request, and Starlink satellites were promptly deployed in Ukraine, quickly integrating into the military action. They provided redundant network support to the Ukrainian government, defense, and key infrastructure departments.

By the end of April 2023, Musk had provided Ukraine with over 10,000 Starlink terminals. Starlink ensured that Ukrainian information communications were never entirely interrupted or paralyzed, even when military infrastructure and heavy weaponry within Ukraine were largely destroyed. The Ukrainian military continued to use Javelin anti-tank missiles, Stinger anti-aircraft missiles, and other individual soldier weapons, along with guerrilla tactics, to inflict significant losses on Russian forces.

Starlink primarily employs a 1.0 version of satellite communication and provides Earth-based Internet communication services to Ukraine through ground gateway stations deployed in neighboring countries such as Poland, Turkey, Lithuania, and Romania. Ukrainian civilians, discovering Russian military operations, used Starlink equipment to transmit intelligence to Ukrainian military personnel, aiding them in using drones for "on-demand" strikes against Russian personnel and equipment, causing considerable casualties.

LEO satellite systems, under the guise of private enterprises and civilian systems, championing the banner of information freedom, and cloaked in the guise of peaceful use of space resources, allow intelligence information to be transmitted unrestricted across national borders, creating information super-vision blind spots for other sovereign nations.

Notably, current weaponry has limited effectiveness in soft-killing LEO satellite systems. Physically destroying in-orbit satellites or ground gateway stations in third-party territories would inevitably violate international law, leading to global public outcry and increasing the risk of hostile nations or international organizations joining the conflict.

The Russia-Ukraine conflict is a classic example of the US and Western countries using LEO communication satellites to conduct a new type of proxy war against Russia. Western strategists have vividly summarized this new proxy war model as "Starlink + Katyusha," representing various indirect and support-ive means in proxy wars, such as information warfare, psychological warfare, and financial sanctions.

First, "Starlink + Katyusha" ensures the smooth flow of military, political orders, and intelligence during war. With "Starlink," Ukrainian command departments not only obtained a wealth of high-value Russian military battle-field situation and military personnel intelligence from NATO but also directed frontline troops for precise strikes against Russian forces in real-time. Addition-ally, frontline Ukrainian troops reported battle situations and requested support from their command institutions, relying on the stable network communication capability provided by "Starlink."

Second, "Starlink + Katyusha" is reflected in maintaining communication for civilians in the war zone and strengthening the international information war. In the early city battles of the Russia-Ukraine conflict, Ukrainian civilians restored traditional mobile communication networks through "Starlink," transmitting a large number of images and text intelligence about Russian forces via mobile phones. This real-time battlefield live broadcast based on civilian perspectives not only garnered extensive international sympathy for Ukraine but also provided material for NATO countries to launch an international information war.

Last, "Starlink + Katyusha" is also evident in supporting unmanned cluster warfare, although this was not fully demonstrated in the Russia-Ukraine conflict. Ukraine mainly used Turkish-made TB-2 drones, which lack satellite communication capabilities and are of low intelligence. Therefore, "Starlink" mainly played a role in facilitating communication between the ground control cabins of drones and frontline troops.

In the foreseeable future, as a new generation of drones embedded with AI mechanisms enters production and spreads to smaller countries, new LEO communication satellites that integrate intelligence, communication, and partial command functions will undoubtedly play a significant role in new proxy wars.

6.3.2 "Integrated" Warfare of the Future

The application of Starlink in the Russia-Ukraine conflict highlights the potential of communication satellites in military applications.

Regarding communication, satellites, with their low-latency, high-speed, and reliable global communication capabilities, can provide beyond-visual-range connectivity for mobile troops, control remote sensors and aircraft, or transmit real-time battlefield information, breaking down the barriers between sensors and attack platforms. In wartime, ground communication infrastructure might be destroyed; LEO satellite constellations offer large communication bandwidth, high efficiency, fast speed, and short delay. They can provide high-quality communication services for drones, aircraft, helicopters, and other mobile equipment, freeing them from land-based communication system constraints. They are no longer affected by terrain and weather conditions such as mountains, oceans, polar regions, or adverse electromagnetic environments, enhancing their resistance to interference and anti-hijacking capabilities.

As the networking speed of communication satellites accelerates, the future 6G network based on satellite communication will reduce latency and provide high-throughput, global coverage, and high-speed satellite Internet beams without any blind spots. Thus, when other connection methods are disabled or interrupted, military personnel can access secure satellite networks to receive up-to-date information about ground situations and enemy and friendly forces. This information can then be seamlessly transmitted to the frontlines or back to headquarters thousands of miles away, enabling commanders to make wise decisions and potentially turn the tide of conflict.

In survey, LEO satellite communication can achieve inter-satellite communication links. Taking Starlink as an example, its constellation layout ensures that several satellites pass overhead in each region every short interval, allowing multiple satellites equipped with reconnaissance payloads to cooperate for all-weather surveillance of specific targets. They can also integrate with ground stations, early warning aircraft, reconnaissance planes, etc., to form an in-depth reconnaissance system that spans air, space, and ground. In November 2019, the US DARPA announced the creation of the "Blackjack" system to monitor globally using LEO satellites. Commercial LEO satellites could become an important platform for the "Blackjack" system. Utilizing commercial LEO satellites carrying reconnaissance and surveillance payloads will further enhance the US military's reconnaissance and surveillance technological advantages.

Accurate navigation on the battlefield is also crucial. For example, once Starlink successfully networks, it could further enhance the accuracy of the US GPS. Starlink not only has a transmission rate of up to 1 Gbps but can also achieve compatibility with GPS signals through software upgrades. Enhancing navigation and positioning capabilities will strengthen the advantages of multi-domain joint operations. The premise of "Multi-domain Warfare" is to use strong navigation, positioning, and information communication capabilities to link tanks, infantry fighting vehicles, self-propelled artillery, helicopters, etc., into an interconnected and communicative whole. This will enhance the "all-domain maneuver" and "cross-domain coordination" capabilities in military operations, strengthening the advantages of multi-domain operational capabilities.

Satellite communication introduces a novel integrated combat mode, poised to further elevate the future of warfare. Integration lies at the heart of this evolution, merging multiple combat domains, armed services, weapon platforms, and information sources into a cohesive, efficient military operation system. This integration enables armed forces to make decisions more rapidly, flexibly, and accurately, achieving highly efficient coordination in joint operations. Indeed, the structure of future wars will become increasingly complex, shifting focus from large-caliber ammunition and long-range missiles toward high-tech, information-driven, intelligent warfare.

6.3.3 Starlink Will Transform Future Warfare Modes

Though only partially and initially constructed, Starlink has already demonstrated significant value in warfare. In the near future, as Starlink completes its deployment, creating a satellite communication network enveloping the Earth in LEO, it will transform the modalities of modern warfare.

On December 1, 2022, the FCC approved SpaceX to launch 7,500 second-generation "Starlink" satellites, allowing SpaceX to commence the deployment of the "Starlink" system while also reviewing SpaceX's proposal to launch 29,988 satellites. Following this, on December 3, SpaceX unveiled the "Starshield" scheme, which relies on the second-generation "Starlink" satellites for its functionality and carries certain militarized characteristics. In essence, "Starshield" is a military version of the "Starlink" constellation.

Given the US military's high dependency on GSO satellites for communication, a military system heavily reliant on Starlink implies an aversion to losing high-value satellites. Consequently, the US military must account for Starlink's risks and contingencies. To mitigate the risk of Starlink becoming a target, the US military has proposed the concept of "mosaic warfare," which involves using low-value, networked, neuron-like systems to replace high-value weapons. Even if the enemy destroys some nodes, the remaining nodes can still form a formidable capability. SpaceX's "Falcon" rockets, with their rapid deployment capabilities, can swiftly deploy a large number of "Starlink" satellites in response to battlefield needs, thus constituting a "mosaic warfare" in space. In the future, SpaceX may develop reconnaissance and strike-specific satellite platforms based on the US military's requirements, leveraging the existing "Starlink" platform through external collaboration and payload integration. The collaboration between SpaceX and the US military is expected to deepen and expand, with space military application technologies and platforms becoming a significant direction for future research and development. The militarization of "Starlink" will have the following seven impacts on future warfare:

1. **Complex electromagnetic space warfare environment.** The 2.0 version of Starlink satellites, as per SpaceX's plan, will distribute nearly 30,000 satellites at altitudes of 328 to 614 km, offering higher bandwidth and lower latency. These satellites will be equipped with specialized surveillance, navigation, and electronic jamming devices, expanding their application in electromagnetic space warfare. The increase in space electromagnetic equipment will complicate the environment, potentially becoming a significant aspect of future space confrontations.

2. **Enhanced redundant network communication capability.** Deep integration of the US military with Starlink and other satellite communication networks will provide redundant communication and command capabilities, enhancing survivability in complex electromagnetic environments. The US Air Force has tested the application of Starlink satellite terminals on military refueling/transport aircraft platforms since 2018. The US Army signed a three-year cooperative research and testing agreement with SpaceX to study Starlink's cross-network data transmission capabilities. The 2020 Convergence Project exercise emphasized the application of commercial satellites in network communication and ground remote sensing, aiming to reduce the sensor-to-shooter time from 20 minutes to under 20 seconds. In March 2022, the US Air Force conducted a high-speed communication test at Hill Air Force Base involving the agile combat deployment of the F-35A Lightning II stealth fighter, where Starlink played a key role.

3. **Enhanced reconnaissance and surveillance capabilities.** In August 2017, when SpaceX planned to expand the Starlink trademark to satellite imaging and remote sensing, it indicated the potential for 24-hour uninterrupted optical monitoring of specific areas using Starlink satellites. Combining Starlink's optical observation equipment with AI can enable automatic identification and tracking of targets. The image recognition system, with a high true target recognition rate and strong anti-interference ability, can conduct real-time surveillance and moving target tracking. Combined with ground stations in the US, Starlink can build an integrated monitoring system, significantly enhancing dynamic perception abilities globally and on regional battlefields.

4. **Improved missile warning capabilities.** Starlink satellites, capable of emitting omnidirectional beams, can not only perform telemetry, tracking, and control of spacecraft but also track missile launches, forming a high-precision system for the calculation, simulation, and prediction of missiles. Therefore, Starlink holds significant importance in missile early warning. The best method to intercept ballistic missiles is before launch or in the early flight stage. If part of the Starlink constellation is transformed into missile warning satellites, real-time monitoring of ground missile bases can be achieved. Sharing missile launch information with the US military and its allies' anti-missile systems, such as THAAD and Aegis, will significantly increase interception success rates.

5. **Advanced precision navigation and positioning capabilities.** Starlink satellites' navigation and positioning function, calculated by the Ohio State University Multi-modal Assured Navigation Autonomous Vehicle Research Center, can achieve Earth surface positioning accuracy of 8 m. Starlink could reach

GPS navigation satellite accuracy of 0.3 m or higher if connected to thousands of satellites. However, the anti-interference capability of Starlink's Doppler-effect-based positioning is not up to military requirements. It is more suitable as an alternative navigation positioning method when the GPS is jammed. With the realization and maturity of quantum communication satellite technology, the next generation of Starlink based on quantum communication will play a more significant role in the military field.

6. **Upgraded space electronic warfare capability.** Starlink satellites can enhance monitoring and jamming capabilities against terrestrial and space targets by carrying electronic warfare equipment. The LEO has become a focal point of competition among nations. With the development of various countries' low-orbit satellite projects, the number of electronic warfare equipment in space will increase, driving the growth of countermeasure equipment and upgrading space confrontation technology. The US military is currently developing space electronic warfare equipment specifically for ground radio frequency source monitoring and integrating such equipment with Starlink is quite feasible.

7. **New communication relay.** Currently, the US military cannot directly integrate Starlink terminals into the F-35 fighter jet. When the F-35 needs to execute highly stealthy combat missions without carrying electronic pods, Starlink terminals can be installed on drones and share data with F-35 and F-22 through military data links to create an "Advanced Tactical Unmanned Relay Platform." Since Starlink can transmit data to the F-35, it can also transmit data to other US military aircraft, especially the B-2 and B-21 stealth bombers. Once integrated with the Starlink system, their stealth and combat capabilities will be greatly enhanced. In 2019, the US military used Starlink for unmanned command testing, proving it could overcome drone communication bottle-necks and enable operators to command many drones simultaneously for group military actions. This means that the integrated communication system based on Starlink can provide operational support for drones, unmanned boats, and various other advanced intelligent military equipment.

Starlink is not only rewriting the history of human communication and triggering a new business revolution but is also undoubtedly driving military transformation. Future wars are expected to evolve toward unmanned, intelligent, communication-based, and precision-oriented directions.

6.4 Interstellar Communications: Making Space Travel a Reality

Humanity is on the brink of achieving its grandest ambitions in the current era of exponential growth. Beyond exploring the depths of our world, the far reaches of deep space beckon as the next frontier. From lunar landings to Martian explorations and Pluto's mysteries, the secrets of our solar system are gradually unraveling, making the vast cosmos less enigmatic.

Moreover, pioneers are creating new realms in space, expanding humanity's dimensions. Richard Branson's Virgin Galactic has realized space travel in a 90-minute journey, while Elon Musk's SpaceX has already signed contracts for lunar orbit flights with space tourists. Once confined to science fiction, Space travel is now an approaching reality. However, establishing interstellar communication is a prerequisite before humanity embarks on space travel or colonization.

6.4.1 From the Age of Sail to the Space Age

Human civilization, starting with the Sumerian civilization, has been predominantly land-based. This era saw the rise of empires like Babylon, Rome, Byzantium, Arabia, Han, Tang, and Mongol, among others, and civilization's cradle. While notable maritime civilizations like the Aegean existed, land-based civilizations were mainstream. The Age of Exploration changed the world.

The opening of new maritime routes established the first global connections across continents and oceans. This broke the relative isolation of continents, integrating the world. Since Columbus discovered the New World, maritime empires like Portugal, Spain, Holland, and France rose, and modern America is a product of European maritime activities.

The Age of Exploration reshaped South America, North America, the Caribbean, South Asia, and Southeast Asia societies, transitioning from tribal societies to modern states. Portuguese and Spanish colonization in the Americas laid the foundations of many contemporary South American and Caribbean nations. The English East India Company's establishment of British India led to today's India.

This era reshaped the world's political map and cultural landscapes and triggered globalization, profoundly impacting today's Internet and telecommunication sectors. The demand for global navigation spurred technological revolutions, impacting the rapid development of industries like Internet connectivity.

However, even as the grandeur of the Age of Exploration continues, it is still confined to Earth, a mere part of the solar system. Among the four terrestrial planets—Mercury, Mars, Venus, and Earth—Earth is the largest but dwarfs compared to Jupiter, Saturn, Uranus, and Neptune. Against this backdrop, venturing into space to explore a "new world" seems inevitable. On April 12, 1961, the world's first manned spacecraft launched from the Soviet Union, with Yuri Gagarin becoming the first human in space, ushering in a new era of space exploration.

Amid the US-Soviet rivalry, significant milestones followed: the first female astronaut in 1963, the first spacewalk in 1965, the first space docking in 1966, and the first moon landing by Apollo 11 in 1969.

Subsequently, humanity sought to establish a long-term presence in space—space stations. The Soviet Union launched the world's first manned space station, Salyut 1, in 1971, followed by more until the second-generation Salyut 6 and 7. In 1973, the US launched the Skylab space station, followed by the Mir space station in 1986, which recorded continuous human habitation in space.

From the Age of Sail to the Space Age, 60 years of human spaceflight have paved the way for common people to dream of space travel.

6.4.2 Space Travel: No Longer a Distant Dream

Space travel is heating up. Jeff Bezos of Amazon, Richard Branson of Virgin, and Elon Musk of Tesla have all established their own space tourism companies. These private enterprises are striving to create a new business model for space travel. Musk even boldly stated at the Royal Aeronautical Society in London that within the next 15 to 20 years, SpaceX plans to transport 80,000 people to colonize Mars.

As early as 2018, SpaceX announced that Japanese billionaire and Zozotown founder Yusaku Maezawa would be the first space tourist to fly around the moon with SpaceX. Maezawa has already paid for the entire journey, including the travel costs of eight crew members he plans to gift the trip to. In February 2021, SpaceX unveiled its first private space travel mission, "Inspiration4," to be executed by the Falcon 9 carrying the Dragon spacecraft.

On April 8, 2022, SpaceX completed its first all-private manned space flight. This flight didn't include any active professional astronauts, with ages ranging from 50 to 71. The mission commander was retired NASA astronaut Mike

Lopez-Alegria. Born in Madrid, Spain, on May 30, 1958, Mike Lopez-Alegria is a former NASA astronaut and test pilot, now serving as Axiom's Vice President of Business Development. After high school, he joined the US Navy, earning a bachelor's degree in Systems Engineering and a master's degree in Aeronautical Engineering. He has flown three Space Shuttle missions (STS-73, STS-92, STS-113) and one mission to the International Space Station aboard Soyuz TMA-9. He holds the record for the second-longest duration of spacewalks and is the fifth-longest space flier among US astronauts.

The other three members are Axiom Space customers: pilot Larry Connor, Canadian mission specialist Mark Pathy, and Israeli mission specialist Eytan Stibbe. They each paid $55 million (approximately CNY 350 million) for their trip to the space station. Commercial companies provided the rocket and spacecraft they all traveled on.

NASA revealed that during their time on the space station, each person's daily food costs would be $2,000, with additional transportation and supply costs ranging from $88,000 to $164,000 per day per person. Each commercial launch also included the delivery of supplies for astronauts working on the space station, costing $5.2 million, borne by commercial travelers. Additionally, they had to pay NASA $4.8 million for mission planning and support fees.

Besides Musk's SpaceX, Jeff Bezos' Blue Origin and Richard Branson's Virgin Galactic are also vigorously developing in the commercial aviation race. On July 11, 2021, 71-year-old Virgin Galactic founder Branson flew to space, approximately 86 km from Earth, aboard his developed White Knight spacecraft, safely returning to the desert in New Mexico. Branson's space journey bolstered Virgin Galactic and marked a milestone in commercial aviation.

6.4.3 Constructing the Interstellar Communication Network

As commercial aerospace ventures start offering space travel services, the much-anticipated space age for humanity is on the brink of reality. Although mankind is increasingly close to venturing into space for travel and even colonization, interstellar communication remains a crucial issue to be resolved beforehand. Maintaining communication with Earth during interstellar travel, or communication between stars, is vital and immensely challenging.

Spacecraft cannot be like kites severed from their strings; they must maintain contact with Earth. No matter the distance traveled, spacecraft must occasionally send back information to Earth, even if it's just a simple message of "all is well."

However, achieving this is a large feat. Interstellar communication's complexity primarily stems from the attenuation of radio waves during transmission. In this process, the energy density diminishes inversely with the square of the distance. Simply put, as the spacecraft's distance from Earth increases, the signal's energy weakens, and interference from thermal noise around Earth makes reliable communication exceedingly difficult.

Take the Voyager 1, the fastest and farthest spacecraft from Earth, as an example. It is currently about 17 light-hours from Earth, meaning signals from Voyager 1 take 17 hours to reach our planet. Such vast distances present significant challenges to communication signals. As the signal travels farther, its energy gradually diminishes, and it is considerably weak upon reaching Earth.

To address this, scientists must consider how to maintain reliable communication over extreme distances when designing spacecraft. Despite Voyager 1 being equipped with a relatively large 3.8 m diameter antenna and operating at a power of only 23 W and a communication frequency of 8 GHz, the signal's energy still sharply decreases during transmission. A gigantic 37 m diameter antenna is required to receive such a signal on Earth. Even so, the communication rate remains limited to about 1.4 kbps due to signal attenuation. To put this into perspective, the early Internet dial-up speed at home was about 54 kbps, and transmitting a photo with today's smartphone would take at least 2 hours.

As Voyager 1 continues its journey beyond the solar system, it won't be long before it loses contact with Earth. NASA has a Deep Space Network with antennae measuring 70 m in diameter. The larger the antenna, the greater the signal energy received and the farther the communication distance. However, even with more advanced coding and data compression technologies in the future, the maximum communication distance supported will still fall far short of the needs of interstellar travel.

The key to solving this problem lies in energy. As depicted in the sci-fi novel *The Three-Body Problem*, even though the Red Coast Base had a huge antenna, Ye Wenjie still needed to use the sun as a signal amplifier to establish communication with the Trisolarans. As a celestial body, the sun has immense mass and energy, which can be used to amplify and propagate signals.

Moreover, theoretically, quantum entanglement communication could break the limitations of traditional communication methods, achieving faster-than-light communication. Quantum entanglement is a fascinating quantum phenomenon where the states of two or more particles are interrelated, regardless of the distance between them. This interrelated state can transmit information instantaneously, enabling real-time communication even in the farthest corners of the universe.

However, this theory still requires further research and practical validation to confirm its feasibility. Scientists are tirelessly working toward using quantum entanglement communication to solve future long-distance communication challenges, breaking space and time constraints to enable further exploration and discovery. If humanity truly manages to achieve quantum communication, it would solve the communication issues associated with interstellar travel, allowing human footprints to spread across more planets.

6.5   When Satellites Become Space Debris

As more satellites are launched into space, humanity's pace of space exploration accelerates. However, this brings a problem: space debris. Like Earth, space has a limited environmental capacity—there's a threshold to utilizing space resources governed by space capacity, rules, and order. Today, one of the most severe environmental issues arising from space exploration is space debris, or useless man-made objects orbiting Earth, predominantly fragments of artificial satellites.

Increasing space debris poses new challenges to space exploration, heightening the risk of collisions with space stations and manned spacecraft. Concurrently, this challenge opens opportunities for developing space debris removal technologies and the emergence of a new space debris cleanup industry.

6.5.1 The Growing Problem of Space Debris

As long as humans continue to launch satellites, space debris is inevitable.

Particularly in LEO, which can be understood as a broad lane close to Earth. Most human-launched objects, including space stations, satellites, probes, and spacecraft, cluster in LEO, generating debris predominantly distributed in this orbit.

Recently, due to technological advancements and cost reductions, entering space has become increasingly accessible. Traditional radar satellites are massive, weighing tons and costing hundreds of millions of dollars to launch, thus only affordable for governments. However, satellite launches are gradually becoming commercialized, and satellites are getting smaller. Stanford University's CubeSat is only 10 cm cubed, while Planet's Dove satellites are the size of a shoebox.

Moreover, rockets can carry more satellites. In 2021, SpaceX set a record by launching 143 satellites in a single mission.

With the rapid development of the satellite communication industry, the number of man-made objects in space is surging. Before 2012, fewer than 200 objects were launched into space annually. By 2020, this number skyrocketed to 1,200 and approximately 1,800 in 2021.

Currently, over ten thousand satellites orbit Earth in LEO. With plans to launch even more satellites, LEO is becoming increasingly congested, and the quantity of space debris is significant.

Space debris encompasses discarded rocket boosters, retired satellites, various parts of spacecraft (like screws, nuts, and bolts), and smaller debris, such as paint flakes peeled off spacecraft exteriors due to intense ultraviolet radiation. Due to the vast number of fragments and their small size, making them untraceable, no one knows the exact quantity of space debris.

According to a report by NASA's Orbital Debris Program Office, as of 2018, the total number of satellite fragments and rocket wreckage in near-Earth space had reached 14,357. Research by space debris expert Donald Kessler suggests that, at the current rate of increase, space debris will reach a critical value for a chain collision effect within about 70 years, rendering near-Earth space unusable.

Megan Donahue, a professor of astrophysics at Michigan State University, stated, "We are surrounded by something akin to a landfill." Just over half a century ago, in 1957, there was only one artificial satellite in space—the Soviet Union's Sputnik 1, weighing 83.6 kg and about the size of a metal ball with a half-meter diameter, containing a radio transmitter.

The issue of space debris must be addressed. Collisions with even a few centimeters long space debris can severely damage satellites or spacecraft. Rocket launches sometimes have to be delayed to prevent collisions with small metal fragments floating around Earth. This is a major challenge for humanity's largest "spaceship"—the International Space Station. Scientists regularly need to elevate or descend the station to a safe location, or else it might become space debris itself, posing a life-threatening risk to astronauts aboard.

The "attack principle" of space debris is akin to bullets or missiles but even faster, as objects in Earth orbit generally travel at 27,000 km per hour. Such collisions are usually not head-on, but the average impact speed still reaches 36,000 km per hour.

For instance, an aluminum ball just 1 cm in diameter, roughly the size of a standard marble, can have the impact energy equivalent to a cricket ball traveling at 2,600 km per hour, or a car at 96 km per hour. No object can withstand such an impact.

Furthermore, space debris also threatens Earth's safety. Each year, about 400 pieces of space debris re-enter the atmosphere, some of which are not completely incinerated and pose a serious safety threat when they fall to the ground. On March 29, 2007, the Associated Press reported that on the evening of the 27th, a commercial airliner flying from Santiago, Chile, to Auckland, New Zealand, narrowly missed being struck by falling space debris while flying over the South Pacific Ocean.

Dense and massive debris, rocket remnants, and space lifestyle waste all pose direct or potential threats to spacecraft operation and further space exploration. The uncontrolled floating of vast amounts of space debris disrupts the ecological balance of the space environment. If this breaches the self-regulatory baseline of the space environment, it could lead to an irreversible space environmental crisis.

6.5.2 How Many Ways Are There to Clean Space Debris?

The logic behind generating space debris is simple: while nations compete to launch satellites, no one undertakes to clear them up. Humanity is trying to reduce space debris to prevent scenarios like those in many sci-fi movies where satellite debris triggers accidents.

Knowing the location and speed of space debris in advance, calculating its future trajectory, and notifying spacecraft that might collide with it to change their orbit and avoid impact is only a temporary solution. To eradicate space debris, we need to address the debris itself.

Incineration is a commonly used solution. Most debris slows down in orbit and eventually falls, burning up in the atmosphere over days or years. Some objects are deliberately burned up in Earth's atmosphere. To be safe, their fall is aimed at oceans to avoid densely populated areas. Another solution is to send decommissioned satellites to a higher orbit, the so-called "graveyard orbit," where few spacecraft operate, significantly reducing the likelihood of collisions.

Methods for clearing and recycling space debris include "laser brooms," using light pressure from lasers to push away small fragments or thermal energy to vaporize debris. This method is suitable for space debris 1–10 cm in diameter. Arthur C. Clarke's 1979 sci-fi novel *The Fountains of Paradise* depicts such a scenario—people launch a major cleanup operation using space fortresses equipped with high-energy lasers to sweep the skies and vaporize all debris.

Another method is the "space debris net," made of high-strength fibers to intercept space debris. A US aerospace technology company suggests developing a "Space Shepherd" device to methodically clear space debris. The device, mounted on solar-powered spacecraft, releases small spacecraft called "shepherds" near debris. The shepherds orbit the debris to find docking points and, once connected, drag the debris back into the atmosphere. Each mission can capture space debris several times its own mass, akin to herding sheep with a shepherd dog.

This method is like attaching "wings" to space debris. In March 2010, scientists at the University of Surrey unveiled the "Cube Sail," a 5 × 5 m plastic film deployed in space. Launched separately and guided by ground personnel, it uses its propulsion to approach space debris. Once docked, it unfurls the sail, pulling the debris out of orbit to fall to Earth.

"Mechanical janitors" involve maneuverable satellites with mechanical arms to capture space debris. Wang Jinkang's sci-fi short story *Space Sweepers* also depicts astronauts piloting "space sweeper vehicles" to collect space debris. Frequent human space activities generate increasing amounts of space debris. Safe space flight could become impossible without proper management, making space sweepers an inevitable new profession.

Some space environmentalists think further ahead. Nicholas Johnson, the chief scientist for orbital debris at the Johnson Space Center, believes that space debris around the moon must be considered before returning there. These fragments, remnants of past lunar exploration, threaten future unmanned missions and moon landings. Due to the moon's uneven gravitational field, this debris could escape its orbit and strike the lunar surface at 5,000 miles per hour. This not only threatens the safety of astronauts but could also damage historically significant Apollo landing sites.

6.5.3 The Rise of the Space Debris Cleanup Industry

The burgeoning space debris cleanup industry is evolving alongside advancements in space debris cleanup technology. In recent years, enterprises dedicated to this endeavor have emerged steadily.

In 2019, Clear Space, a Swiss aerospace startup and a leading space debris cleanup technology developer signed a contract exceeding $100 million with the European Space Agency for space debris retrieval. By 2025, Clear Space might become the world's first enterprise specialized in space debris cleanup.

The Clear Space-1 mission will deploy a robotic satellite equipped with four mechanical arms to clean up space debris. As the satellite enters orbit, sensors detect and approach space junk, and then the mechanical arms will envelop the debris, causing it to fall toward Earth. Both the robot and debris will burn up due to frictional heat as they enter the atmosphere.

Japan is also making strides in the space debris cleanup industry. Astroscale, a space debris cleanup company, has developed a method involving launching a robotic satellite that uses an adhesive to stick to space junk and then return to the atmosphere. Astroscale has secured an investment of JPY 21 billion, aiming to commercialize this service around 2023. In March 2021, Astroscale's debris-recovery satellite ELSA-d was launched aboard a Russian Soyuz rocket in LEO.

Russian aerospace company StartRocket is developing a satellite that utilizes a "foam fragment catcher" technology to recover space debris, aiming to launch as early as 2023. This cylindrical satellite releases a sticky polymer foam in areas heavily littered with space debris, capturing the fragments and then propelling them into Earth's atmosphere to burn up from frictional heat.

EOS, an Australian technology company, announced that after seven years of research and development, they have successfully created a powerful laser capable of knocking dangerous space debris from Earth's orbit. This laser accurately tracks and hits space debris in Earth's orbit. If operational, this laser could make the cosmos safer than it is now.

According to a report by 9News, this system actually consists of two types of laser beams. The first, a bright orange laser beam, is responsible for targeting specific space debris; the second, for enhancing the accuracy of the laser, is significantly more powerful than the first. The latter laser is shot out of orbit into deeper space. After mapping the atmosphere, the system, updating its map hundreds of times per second, adjusts the ground-based laser beams, thus perfectly targeting space debris.

South Korea is also considering developing related technologies. According to the South Korean government's 2021 Space Hazard Response Implementation Plan, the government will invest approximately KRW 13 billion in technology development to address the risks of space object collisions and space hazards. Additionally, South Korea plans to join international organizations like the Inter-Agency Space Debris Coordination Committee and the International

Civil Aviation Organization, which are focused on developing space debris cleanup technologies.

It is foreseeable that utilizing, and manufacturing space resources will bring substantial profits and redefine humanity's future, but this hinges on a clean and safe space environment. Nowadays, the awareness of environmental protection extends beyond the atmosphere. Indeed, humanity can explore space only under the dual constraints of evolving international law and self-regulation of ethics. If there is a lack of international consensus in areas such as space debris cleanup, space resource extraction, and basic conduct in outer space, the bright future envisioned by humanity may be compromised.

6.6   The 6G Era: The Need for Space Law

Today, humanity is expanding into space at an unprecedented pace, deeply developing and utilizing space and its resources. Amid escalating international strategic competition, space has become a new focus for the strategic games of major powers and a new chip for strategic checks and balances. To seize the high ground in strategic competition and control the initiative in space development, major countries worldwide are adjusting their space strategies, driving a new development trend in international space strategic competition. Whether it's countries vying for orbital space resources or the US SpaceX's satellites forcing the Chinese space station to take emergency collision avoidance maneuvers twice, it's evident that the winds of space competition are gaining strength.

The strategic and security games among major powers extend from traditional domains to space. As space competition intensifies, the security risks of disordered space competition become increasingly prominent. However, neither internationally nor among major countries have established specific protective measures for space.

6.6.1   Intensifying Space Competition

In the past, a country's space capabilities and position in international relations were the main factors determining the space landscape. Space capabilities dictated the scope and depth of activities in space, and international status laid the political foundation in the space arena. However, space technology is increasingly becoming the foundation supporting space strategy. Space technological capability gradually surpasses international status as the decisive factor shaping the global space competition landscape.

Thus, to gain an edge in the fierce space race, countries have successively introduced ambitious space strategies, development plans, exploration concepts, and policy measures, ramping up investment in space technology R&D.

The US government's National Space Strategy emphasizes the "America First" principle, highlighting US leadership in space and a hegemonic strategy of seeking peace through strength. To effectively implement the space strategy and ensure that "America's dominance in space is never questioned or threatened," the US has virtually reshaped its entire space domain's management system and functional system. Russia, Western European countries, Canada, Japan, and South Korea have also formulated or announced space strategy plans named as policies, guidelines, frameworks, basic laws, and plans. This has directly triggered rapid development in high-tech sectors such as space, information, new materials, and new energy. A new round of high-tech is ready to emerge.

Globally, the US is currently the country with the strongest comprehensive space capability. As of December 2020, the US accounted for more than 56% of the total number of satellites in orbit worldwide, ranking first globally and exceeding the total number of satellites in orbit of all other countries combined. It is 4.6 times that of China, which ranks second, and 10.7 times that of Russia, which ranks third. The US is also the only country to have achieved manned moon landings, sent artificial probes out of the solar system, and systematically equipped anti-satellite weapons.

Despite having an absolute advantage, the US has continued its pace in space strategy. On December 20, 2019, the US passed the 2020 National Defense Authorization Act, authorizing the formation of an independent military branch, the Space Force, and approved a $32 million budget for creating the Space Force headquarters. In the 2021 fiscal year defense budget request, the Space Force's share significantly increased to $15.4 billion.

In the US, aerospace technology is not exclusively governmental; even private enterprises can participate in aerospace technology. As early as January 2015, SpaceX CEO Elon Musk announced his plan to launch about 12,000 communication satellites into space orbit with the "Starlink" project, forming a global, efficient satellite communication network to provide consumers worldwide with cheap, fast broadband Internet services. Today, SpaceX has become the company with the most satellites worldwide.

Certainly, as its own strength develops, China's attractiveness in the field of space is also growing. China's aerospace has made great breakthroughs in rockets, manned spaceflight, satellites, and deep space exploration.

In terms of rockets, China has a complete lineup from series to bundle, from single satellite launches to multiple satellite launches, and from launching satellites to launching manned spacecraft and space probes. In manned spaceflight, from the Shenzhou series to the Tiangong series, China has mastered the three basic technologies of manned spaceflight: human space travel, extravehicular activity, and rendezvous and docking, and has successfully launched the first set of space station modules.

In terms of satellites, China's satellite industry has partially entered the ranks of advanced countries. Particularly notable is the BeiDou Navigation Satellites System; in 2020, the BeiDou-3 system was completed and began providing global services, not limited to the traditional navigation functions of GPS satellites. In lunar and deep space exploration, the "Chang'e-5" mission achieved China's first lunar surface sampling, lunar takeoff, lunar orbit rendezvous and docking, and sample return, marking a successful conclusion to the "orbit, land, and return" steps of the lunar exploration program.

The "Tianwen-1" Mars probe has reached Mars, and the "Zhurong" Mars rover has successfully landed on the Martian surface. The next step is to implement a Mars sample return mission in 2028, and before 2025, conduct sample return missions from near-Earth asteroids and orbiting exploration of main belt comets, and in 2029, implement exploration missions to the Jupiter system and interplanetary traversal.

Currently, competition for Earth's resources is essentially saturated, and it's unlikely that humanity will again engage in large-scale wars over territorial disputes. Space competition is increasingly becoming a new focus for major powers' strategic games and a new chip for strategic checks and balances. Whoever gains dominance in space first will hold the initiative in the next era of technology, economy, military, and other fields. In this context, the two most influential countries in the world, China and the US, are intensifying their competition in the field of space.

6.6.2 Disorderly Space Competition

Certainly, not all competition is orderly, especially in the unregulated realm of space. As the strategic competition among major powers intensifies, the relationship between cooperation and competition in the space field becomes increasingly complex.

On the one hand, as the competition and gamesmanship among major powers in space intensify, their struggle for the international space market becomes more evident, further promoting a situation where "competition supersedes cooperation" and increasing the risk of space conflicts. For instance, on December 3, 2021, China submitted a note through the United Nations Committee on the Peaceful Uses of Outer Space to the United Nations Secretary-General, revealing the insider details of how the US SpaceX's Starlink satellites twice threatened the Chinese space station: once by precisely adjusting their orbital height and then approaching a few days later, and another time by directly charging toward the Chinese space station, a deeply concerning scenario.

In the first instance, the Starlink-1095 satellite, which had been operating stably at an average height of about 555 km, began to continuously lower its orbit in mid-May, reaching a height of 382 km on June 24 and maintaining its operation, just close enough to the 390 km orbit of the Chinese space station, seriously violating the ITU rules and creating a potential disaster.

In the second instance, on October 21, another Starlink satellite, 2305, again maneuvered continuously, crossing the orbital height of the Chinese space station and charging straight toward it without any pause, coming as close as just 4 km from the space station. Given the satellite's speed of 7 to 8 km per second, there was less than a second to avoid a collision.

Furthermore, it is important to note that satellite orbits and frequencies are strategic resources. A LEO satellite is part of a network of thousands of small satellites, which do not end their service life all at once. When a few satellites are damaged or end their lifespan, new satellites are launched to replace them rather than relinquishing the entire orbital resource. LEO satellites are limited and vary in quality, with different frequencies experiencing different transmission losses.

Internationally, satellite frequencies and orbital resources are primarily allocated through a "first-come, first-served" method of appropriation. As orbital resources are an important threshold, they inevitably intensify the rivalry among major powers, bringing geopolitics into outer space. This leads to developed countries launching satellites first, using them preferentially, and crowding the quality orbits and frequencies, leaving latecomer countries in a passive position with diminishing space for space development.

On the other hand, as the number of human space explorations increases, so does the production of space debris, a cost of humanity's disorderly competition that space has to bear. Although it poses no immediate risk or harm to human life, in the long term, just as industrialization has damaged the Earth's environment, humanity will ultimately have to pay the price for its actions.

6.6.3 How Do Earth Laws Govern Space?

In fact, life and production on Earth are now connected with outer space, as satellites are most commonly used for various communications—be it broadcasting television, relaying phone calls or Internet, connecting IoT devices, or even smart home devices. Everyone uses location navigation, and much of our information about climate change comes from satellites because they can observe the Earth comprehensively.

Outer space has become a part of human life and a part of the continuous development of human society's infrastructure. The impact of outer space on life on Earth is significant. However, neither internationally nor among major countries have specific protective measures for space been established to date. The basic legal framework for space exploration dates back to the Outer Space Treaty drafted by the United Nations in 1967. Although most countries have adopted the treaty, it is essentially a relic of the Cold War era.

Additionally, since the mid-20th century, countries have begun space exploration activities and have subsequently enacted five international treaties, including the 1963 Declaration of Legal Principles Governing the Activities of States in the Exploration and Use of Outer Space, the 1973 Convention on the Registration of Objects Launched into Outer Space, and the 1979 Moon Treaty, collectively constructing the basic rules for space activities. However, limited by the legislative period and the technological level at the time, the content of the treaties mainly focuses on the denuclearization and non-military use of outer space, with less involvement in issues such as the development and commercial use of space resources.

From the current space exploration rules, international treaties are still the basis of governance. The content could be more specific; member states are limited, enforcement is weak, and only a few international legal rules are clarified. The key factors in current space resource development are not elucidated, and the treaties have no binding force on countries that have not joined. Simultaneously, the current multilateral governance mechanism for space resource development is still incomplete and unclear, and the relevant rules urgently need clarification. Both space powers and non-governmental organizations are preparing for the update and development of the mechanism.

Meanwhile, the unilateral governance steps of space powers are developing rapidly, such as the US and Luxembourg, both of which have passed unilateral legislation to protect the space resource ownership of their enterprises or individuals. However, in the legislative work of the Space Committee, some countries have strongly opposed unilateral legislation, arguing that the practices of these two countries violate international law. They point out that unilateral legislation infringes on the principle that "space resources are the common heritage of mankind"; the "first-come, first-served" principle for determining resource ownership infringes on the interests of latecomer countries; and space powers have not provided international assistance and technological dissemination.

Today, disagreements still exist. In 2017, the Space Committee added related topics, hoping that countries would actively participate in the legislative process, construct a multilateral governance framework, and discuss and formulate rules for space resource development. Undoubtedly, human society needs a clear international governance framework to develop space resources to keep tensions at the lowest level and maintain harmony in space.

Space should not be a place of conflict but can be a platform for international cooperation. Just like Earth's regulations, as the way space is used continues to evolve, the rules regulating space use should also be continually updated. As space lawyer Michelle Hanlon stated, "The current legal norms for space conduct are not sufficient, and more specific international regulations need to be formulated."

For instance, the minimum distance between satellites should be defined. This is crucial to avoid suspicion about the intentions of another satellite approaching one's own. When they are too close, it becomes difficult to discern whether the other party is aiming for surveillance or positioning for an attack. Understanding the rules of how people use space daily is essential so that anomalies can be identified promptly in a crisis.

Humans originated from Earth but will not stop at Earth. With the development of space technology and the increasing number of participants in space exploration, when humanity breaks through the atmospheric barrier and heads toward the cosmos, facing another horizon should be more prudent. Rules are for protection, and protecting space is also about protecting humanity.

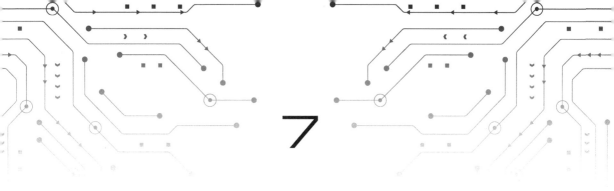

7

THE BIRTH OF
THE STARLINK EMPIRE

7.1 Starlink, the Dominant Force in Satellite Communication

In 2002, Elon Musk founded SpaceX. In 2015, SpaceX launched the "Starlink" project, announcing plans to deploy 12,000 communication satellites into LEO. By March 2018, the US FCC had approved the deployment of 4,425 satellites into orbit. In November of the same year, the FCC further approved an additional 7,518 satellites, paving the way for SpaceX to create a comprehensive space satellite network. In 2022, SpaceX submitted a proposal to the FCC to add 30,000 satellites, bringing the total to approximately 42,000—this increase in satellite density aimed to expand Starlink's coverage and service quality further.

42,000 satellites are an astonishing number. Since the launch of the first spacecraft in 1957, through the US-Soviet space race, the commercial space craze of the 1990s, and the recent surge in Internet service satellites, humanity has only just surpassed the launch of 10,000 satellites. Yet, Musk planned to launch four times that number, a move unprecedented in human space exploration.

According to Musk's vision, Starlink would provide complete global coverage through many LEO satellites. Starlink offers broadband services of at least 1 Gbps, which is five to ten times faster than current home broadband speeds and comparable to the highest speeds of 23 Gbps seen in ultra-high-speed broadband networks. Consumers would only need to purchase SpaceX's terminal equipment and spend a few dozen monthly dollars to enjoy this network.

In May 2019, with the launch of the first batch of 60 Starlink satellites, the Starlink project officially commenced operations. Since then, SpaceX has continuously used its self-developed rockets to transport artificial satellites into space. Since 2023, based on its mature satellite manufacturing, rocket launching, and space-based communication capabilities, SpaceX has expanded the in-orbit scale of the "Starlink" constellation. As of December 2023, the "Starlink" constellation has over 5,000 satellites in orbit, accounting for more than 55% of the total number of satellites globally. It is not only the largest satellite constellation in human history but also the absolute dominator in the satellite communication industry.

7.1.1   Unveiling Starlink's Leading Edge

Starlink's ability to continuously launch satellites into space while sustaining the costs of these launches is due to SpaceX's advanced aerospace technology and capacity for industrial mass production.

SpaceX is a reformer in the aerospace field and a pioneer in the engineering, industrialization, and mass production of rockets and satellites. SpaceX has established its own rocket and satellite industry, drawing on standardization, assembly line practices, and cost-efficiency design and production concepts from modern large-scale industrial manufacturing.

According to research by Georgetown Security Studies Review, the cost per kilogram for launching SpaceX's Falcon 9 rocket has decreased from $10,000 in 2009 to $1,520 today. Currently, in the testing phase, the Starship rocket is expected to further reduce costs to $970 per kilogram. So, how does Starlink achieve this?

The key lies in two aspects. First, the partial recovery and reuse of Starlink's rockets significantly reduce launch costs, an important factor that makes the Starlink project commercially viable.

Rocket recovery is easy to say but hard to do. SpaceX once described the difficulty of vertically recovering rockets: recovering on a land platform is like throwing a pencil over the top of the Empire State Building to land it in a shoebox during a storm; while at sea, it's like landing the pencil precisely on a floating eraser, and it mustn't tip over. The biggest challenge in rocket recovery is the need for the engine to restart during the return to decelerate. The rocket's weight decreases as propellant is consumed, and the rocket's attitude is unstable due to air resistance, requiring the engine to have a wide range of thrust adjustment capabilities for precise control of thrust, descent attitude, and landing point.

Despite the immense difficulty, SpaceX has succeeded in rocket recovery, not without numerous failures and recoveries. Unlike the traditional aerospace industry's emphasis on high success rates, SpaceX advocates for multiple trials and errors. Walter Isaacson, in *Elon Musk: Tesla, SpaceX, and the Quest for a Fantastic Future*, mentioned, "Elon adopted an iterative design approach: rapidly building prototypes of rockets and engines, testing them, blowing them up, modifying, and trying again, until finally making something workable. Progress quickly, blow up test pieces, and repeat the process." News of Starlink satellite launches failures, like rocket explosions and satellite losses, is common. Through each success and failure, SpaceX has tried various new technologies, equipment, and materials, accumulating a wealth of experience, lessons, and massive experimental data. These trials have ultimately propelled the rapid technological development of the Starlink system, significantly enhancing engineering and operational capabilities.

On December 21, 2015, the "Falcon 9" placed 11 satellites into LEO and successfully landed its first-stage booster at Cape Canaveral Air Force Station, achieving the first-ever vertical takeoff and landing mode booster recovery for an orbital launch in human spaceflight history. Musk claimed that "Falcon 9" could complete ten launches with almost no refurbishment and up to 100 launches with refurbishment. As of December 23, 2023, Falcon 9's highest reusability record is 19 times. Falcon 9 has also become the world's most frequently used first-stage rocket for SpaceX and globally.

Naturally, reusing recovered rockets significantly reduces launch costs. SpaceX has achieved the recovery and reuse of first-stage rockets and fairings. Excluding insurance and launch site telemetry costs, rocket launch costs are reduced to the "depreciation cost" and maintenance of the first-stage rocket, the cost of the second-stage rocket, the "depreciation cost" of the fairings, and fuel costs. With the fully and rapidly reusable Starship rocket in use, the deployment cost of Starlink satellites will decrease further.

SpaceX has also made many other efforts in rocket recovery and reuse. For example, carbon fiber is strong and lightweight and widely used in aerospace materials. However, when exposed to continuous high-temperature flames, carbon fiber can experience oxygen-rich combustion. Stainless steel, on the other hand, boasts greater strength, can withstand extremely high and low temperatures, and has lower processing and maintenance costs and faster manufacturing speeds. Therefore, SpaceX uses stainless steel as the rocket material, allowing for rapid reuse after rocket landings.

Moreover, to reduce the difficulty and cost of rocket transportation, the various components of SpaceX's Falcon 9 rocket are produced in different factories across the country, each undergoing testing at their respective sites before being transported to the Kennedy Center for final assembly and integration before launch. SpaceX has also established the Starship manufacturing and launch base, Starbase, in Boca Chica, Texas, which came into operation in 2020. Starbase houses the world's tallest rocket launch tower. The completed boosters and spacecraft undergo final integration assembly at the launch tower, significantly reducing the transportation costs of the super-heavy rocket.

The second factor that significantly reduces costs for the Starlink project is SpaceX's ability to mass-produce satellites industrially. Historically, compared to industrially mass-produced items like phones, televisions, and cars, communication satellites have always been a custom-driven industry. In the past, though thousands of satellites were launched, they were not sent up simultaneously, and their designs were not entirely identical; each satellite might have different missions and attributes. Thus, for a long time, satellite development was characterized by customization—each was redefined, redesigned, and reproduced from scratch, lacking industrialization.

However, with the advent of near-Earth satellite communication networks, to achieve global satellite signal coverage, thousands of satellites are required. Therefore, at the current stage of development, those who can industrialize first and improve cost-effectiveness will have a significant advantage in competition. In this process of satellite industrialization, SpaceX has already taken a step ahead and accumulated experience.

Specifically, SpaceX's "star-making" adheres to Musk's consistent style of self-developing core components to reduce costs and increase efficiency. High-throughput communication antennae, inter-satellite laser communication equipment, Hall thrusters, and other components on Starlink satellites are all self-developed and manufactured. In terms of price, the manufacturing cost of the Starlink satellite V2 mini version has been reduced to around $500,000, with further reductions possible as technology improves and production scales up. According to data from March 2022, Starlink's satellite production capacity has reached eight per day.

7.1.2 Expansion of Service Range and Enhancement of Communication Capabilities

Currently, the Starlink system, with its early advantage in global coverage and commercial space-based communication capabilities, has become the world's most competitive provider of space-based network services.

From a business model perspective, Starlink primarily provides broadband access services to users in areas such as remote regions that lack access to high-speed Internet. Currently, Starlink offers four major types of services, with monthly network service fees as the primary profit model, aside from the one-time hardware terminal charges:

1. **Residential service**. This is Starlink's most basic service. After a price adjustment in 2022, the current terminal hardware fee is $599, with a monthly subscription fee of $110. The expected download speed ranges from 50–200 Mbps, and the upload speed from 10–20 Mbps.

2. **Business service**. Launched in February 2022, this premium version targets small business users and super-users, offering more robust and better-performing terminal equipment, including larger rectangular antennae and upgraded routers. It promises round-the-clock priority service with broadband access. The terminal upfront cost is $2,500, with a monthly fee of $500, expected download speeds of 100–350 Mbps, and upload speeds of 10–40 Mbps.

3. **RV service**. Launched in May 2022, it caters to RV users and those who enjoy road trips or camping. The terminal cost is the same as the regular version, with a monthly fee of $135. However, the RV version has a lower priority, with significant speed reductions during peak times (expected download speed dropping from 50–200 Mbps to 5–100 Mbps, and upload speed from 10–20 Mbps to 1–10 Mbps).

4. **Maritime service**. Launched in July 2022, this version has a monthly fee of $5,000 and comes with two sets of terminal equipment, costing $10,000. The expected download speed is 100–350 Mbps, and the upload speed is 20–40 Mbps.

In terms of service range, since 2023, the Starlink system, while steadily dominating the commercial space-based network service market in the Americas, Europe, and Oceania, has been leveraging its early mover advantage to vigorously expand into the Asian and African space-based communication markets.

Specifically, after Japan became the first Asian country to open Starlink services in October 2022, SpaceX further expanded the Asian Starlink market in 2023. In February 2023, SpaceX began providing Starlink network services in the Philippines, making it the first Southeast Asian country with Starlink. SpaceX reported that Philippine Starlink customers could enjoy download speeds of 50–200 Mbps and a 30-day trial service. According to orders, the monthly service fee in the Philippines is $49, with hardware costs of $533. In March, SpaceX established the Starlink Korea subsidiary in Seoul, planning to launch Starlink satellite network services in South Korea by 2024. In June, the Cluny human rights organization reported limited use of Starlink satellite network services in parts of Kayah State, Myanmar. However, due to the high cost of its network services, it has not yet become widespread in Myanmar. In July, Mongolia approved SpaceX's Starlink system to provide network communication services, stating that millions of Mongolian Internet users could access high-speed connections through Starlink. That same month, Malaysia granted a license to Starlink, allowing it to provide network communication services within Malaysia. In November, Malaysian comprehensive telecommunications and digital infrastructure service provider REDtone and SpaceX signed an authorized distributor agreement to provide high-speed Internet services based on the Starlink satellite in various parts of Malaysia. Also in November, Kazakhstan conducted network connection tests using the Starlink system in ten rural schools and achieved the expected results. Based on this, Kazakhstan plans to provide Internet services to 200 rural schools using the Starlink system.

Following the opening of the Asian market, SpaceX successfully broke into the African communications market, progressively advancing the application of Starlink services in Africa. In February 2023, SpaceX officially began providing Starlink services in Nigeria, marking it as the first African country to access Starlink services. SpaceX stated that Starlink services would cover the entire territory of Nigeria, playing an important role in bridging the rural Internet access gap, supporting national governance, and improving public education. The same month, Rwanda launched Starlink services, selecting 500 unconnected schools as service pilot sites, and plans to connect 3,000 unconnected schools to Starlink network services by 2024 through financing from the World Bank and China Export-Import Bank. In June, SpaceX introduced

Starlink services in Mozambique, making it the third African country to use Starlink services. In October, SpaceX launched Starlink services in Zambia.

In addition to expanding its service range, Starlink's communication capabilities are continuously improving, and it is gradually rolling out diversified services to meet market demands. In terms of airborne services, Starlink is currently providing in-flight Internet services for JSX, Hawaiian Airlines, airBaltic, and ZIPAIR. In October 2023, JSX stated that since January, it had equipped all its aircraft with Starlink airborne connectivity systems. As Starlink's first global aviation customer, JSX is providing passengers with free Wi-Fi services. Passenger test data shows that during flight, Starlink's connection speed can reach up to 160 Mbps, sufficient for video calls and streaming services, despite some fluctuations in speed.

In maritime services, SpaceX announced in April 2023 that the Starlink network had achieved communication coverage across global waters. In June, Australian company Speedcast and Lindblad Expeditions signed a multi-year renewal contract, incorporating Starlink services into their communication offerings. In September, SpaceX and Luxembourg-based SES agreed to provide satellite communication network services for maritime users such as cruise operators. In October, Danish conglomerate Maersk collaborated with SpaceX to install Starlink systems on over 330 cargo ships.

Regarding railway services, in May 2023, SpaceX and Brightline rail company reached a cooperation agreement to provide free Starlink network services on ten trains in certain cities in Florida.

7.1.3   Milestone: Starlink Achieves Break-Even

Satellite communication is an industry notorious for its heavy expenditures, but on November 2, 2023, Elon Musk announced via social media that SpaceX's Starlink business had achieved a cash flow breakeven. For SpaceX, this news marks a milestone in progress.

Despite the market's optimism about Starlink, predicting future annual revenues of up to $30 billion, it required a high-risk investment strategy with high upfront costs, long cycles, and rapid deployment, involving an initial investment of at least $10 billion and a long-term investment of $20–$30 billion. For a long time, Starlink was perceived as a "cash flow abyss" for SpaceX. Musk even stated in 2022 that Starlink was still operating at a loss, and his biggest goal for SpaceX was to "avoid bankruptcy."

Therefore, the achievement of breakeven cash flow for Starlink is of significant importance. It signifies not only Starlink's financial sustainability and growth but also its crossing over the abyss of negative cash flow. In the future, SpaceX may no longer need massive financing cash flows as its corporate value is gradually emerging.

A similar situation occurred with Tesla. In 2018, although Tesla had not yet turned a profit, it achieved cash flow breakeven for the first time, with a significant positive net cash flow from operating activities—Tesla generated $1.4 billion in cash flow from operations in the third quarter of 2018, resulting in a free cash flow of $881 million. This marked a crucial turning point for the electric car manufacturer, which had been burning billions of dollars each quarter. According to the annual report, Tesla's cash flow from operations in 2018 was $2.098 billion. In contrast, Tesla's operating cash flow was negative before 2018, and it had substantial capital expenditures. Subsequently, with the continued increase in Model 3 production and sales, Tesla successfully navigated through financial difficulties, layoffs, heavy fines from US regulators, and frequent short-selling by institutions, becoming the leader in the autonomous vehicle industry today.

Currently, financial details about SpaceX are scarce. According to multiple media reports from July to September, SpaceX's revenue doubled to $4.6 billion in 2022; total costs increased by 56% to $5.2 billion; with a first-quarter profit of $55 million and revenue of $1.5 billion. SpaceX projects that its revenue will double to about $8 billion in 2023. It is important to note that Starlink is still in its early stages, with just over 4,400 satellites in operation and slightly over 2 million subscribers. However, a few years from now, once it progresses to the intermediate or advanced stages with 12,000 deployed satellites and over 20 million subscribers across most countries/regions, Starlink's annual revenue is expected to reach $30 billion or more.

Presently, SpaceX has become one of the world's highest-valued unicorns. After completing its second round of equity sales in July 2023, SpaceX's valuation approached $150 billion, indicating an increase from its $137 billion valuation during financing in January 2023. It is foreseeable that, should Starlink independently go public and raise capital from SpaceX in the future, its market value will be immeasurable.

7.2 From Starlink to Starshield: From Civil to Military Use

As the dominant force in the satellite communication industry, the significance of Starlink is indisputable.

We all know that Internet communication has brought tremendous changes to our lives, deeply integrating and transforming various industries, thereby overturning the entire social development model. The commonplace services we use today, like food delivery and ride-hailing, are revolutionary changes brought about by the integration of Internet communication into traditional industries. The importance of Starlink is no less than the changes brought about by Internet communication. Besides its transformative power in civilian domains, its significance in the military sphere is equally disruptive.

7.2.1 Precision Strikes with Starlink

In the military field, the flow and transmission of information are critical. A slight miss can lead to significant errors. However, in past wars, information often struggled to be conveyed quickly and accurately. Without precise information, there can be no precise strikes.

For example, even if radios are equipped with individual soldiers, it's challenging to achieve real-time monitoring or visualization of each soldier's situation. It's even harder to enable direct information communication and command between individual soldiers and neighboring units. As for a soldier controlling a weapon to strike beyond visual range without a massive support system, it was once an unachievable feat. However, the advent of Starlink has made all this possible.

The greatest value of Starlink lies in providing an inexpensive, globally covered wireless communication network, connecting all command systems and combat units in one network. Each combat unit becomes a node, allowing command centers to command combat units in real time via this network, and units can directly share target information data with each other. This brings the possibility of precision strikes on the battlefield.

Take, for example, artillery strikes on a specific area. Traditionally, scouts or satellite photo interpreters would identify the target coordinates and report them to the command post, which would then relay the target information to each artillery position. Each position would calculate the distance to the target, determine the artillery's direction and scale, and then fire upon command.

With all nodes connected by Starlink, it's like bringing everyone into a group chat. Once a scout identifies the enemy artillery position with a drone and sends the GPS information to the group, each artillery position can calculate the firing direction and scale with a single click. Automated artillery systems can adjust the guns automatically, and the command post only needs to order to fire. This is actually a reality on the Russia-Ukraine battlefield.

Ukraine, bolstered by Starlink, has developed the "Kropiva" universal combat system. Its underlying system is essentially an electronic map and communication software. "Kropiva" features short text distribution capabilities, allowing the dissemination of friendly and enemy troop locations and target coordinate information. It can also solve individual computational problems, such as route calculations, coverage area, or shooting corrections, and enables real-time connectivity with higher and lower command levels through the network.

Moreover, this system, developed on the Android platform, can be easily installed on any Android phone or tablet. Then, by connecting to a Starlink ground station, the system enables units, squads, or individual vehicles or artillery to either operate independently based on their location or coordinate with neighboring troops for combined operations.

Starlink has also enhanced unmanned combat weapons, which have already shown their prowess on the Russia-Ukraine battlefield. The major limitation of such weapons in the past was the remote-control distance, dictated by their communication capabilities. In the Russia-Ukraine conflict, Ukraine leveraged Starlink's communication function to enable long-distance remote control of unmanned boats. In August of this year, Ukraine used unmanned boats to attack the Russian military port of Novorossiysk on the Black Sea coast, with one drone boat carrying 450 kg of high explosives and accurately hitting a large Russian landing ship, causing it to take on water on its port side.

It is foreseeable that in the future, Starlink will further integrate with robotic soldiers and unmanned combat aircraft to realize unmanned, precision warfare. It could also be combined with biomimetic robotic insects for precise decapitation strikes at specific targets. The widespread application of Starlink technology will provide reliable remote communication support for these unmanned weapons, enabling them to maintain efficient command and control globally, thereby increasing their effectiveness on the battlefield.

7.2.2 The Growing Military Contracts of Starlink and Starshield

While Starlink continues to establish its presence in the civilian domain, it is also receiving an increasing number of military contracts.

In fact, as early as March 2019, the US military signed a contract worth $28 million with SpaceX to demonstrate and validate military services using Starlink. In November of the same year, during a low-orbit technology verification test, Starlink provided network services with bandwidths up to 610 Mbps for the US military's C-12 aircraft. The same year, DARPA announced the development of the "Blackjack" system, which uses LEO satellites for global surveillance. It is highly probable that Starlink's LEO satellites have become part of the "Blackjack" system's platform.

In 2020, the US Army also signed a cooperative research and development agreement with SpaceX to connect Starlink broadband satellites to military communication networks and integrate the military's procurement strategies based on the assessment results.

Moreover, some of Starlink's launch sites are located within the Vandenberg Air Force Base. The technological verification tests of its interconnection with military aircraft are also classified as military secrets. Throughout Starlink's development, the US military has provided significant funding, supporting the expansion of Starlink's application scenarios to military fields.

On December 3, 2022, SpaceX officially launched the "Star Shield" program on its website. SpaceX defines Star Shield as a satellite constellation serving national security, distinct from Starlink's commercial operations. Star Shield is dedicated to serving US military and governmental departments, primarily offering remote sensing, communication, and payload hosting services. Furthermore, Star Shield employs additional encryption technology on top of Starlink's data encryption services to ensure the security of hosted payload data processing, meeting official requirements.

If Star Shield represents a new constellation, SpaceX would need to apply for additional orbital slots and spectrum resources, which they have not done. In December 2022, the FCC partially approved SpaceX's Starlink Gen2 constellation plan, authorizing three of the nine orbital shells, totaling approximately 7,500 satellites for the BS plan, while delaying the review of SpaceX's use of E-band frequencies and tracking beacons.

The launch of Star Shield almost coincided with the FCC's approval, suggesting that Star Shield is likely the upcoming BS phase of the second-generation Starlink constellation. Born from Starlink, Starshield's formal announcement to serve in defense opens a new chapter in the military application of commercial space.

From Starlink to Starshield, they have become a crucial support for the US defense forces. Once the Starlink and Starshield projects are fully deployed, they will demonstrate immense military value.

First, 42,000 near-Earth orbit satellites in high, medium, and low layers will "enclose" the Earth, forming a dense satellite network, greatly enhancing US military communication capabilities, and improving the accuracy and anti-interference abilities of the US military's navigation and positioning systems. US precision-guided munitions, supported by Starlink satellites, will achieve more accurate strikes, significantly increasing the difficulty of enemy defense.

Take submarine underwater communication as an example. When a submarine dives hundreds of meters deep, the only viable communication method is long-wave communication, which is limited and prone to errors. With Starlink's global coverage, submarines can use specialized communication buoys for complex communications anywhere.

Second, Starlink will establish an all-weather, uninterrupted intelligence reconnaissance network. Especially in wartime, Starlink can significantly enhance the US strategic missile defense early-warning capabilities. In the future, tens of thousands of Starlink satellites constantly traversing all global regions will help the US military achieve near-continuous reconnaissance and surveillance. Every movement and number of any ground troop can be clearly observed, making the battlefield situation unilaterally transparent. For example, on the battlefield, Starlink can use just a portion of its satellites to monitor enemy ground missile bases in real-time, providing early warning as soon as missiles are launched, significantly improving US missile defense success rates. This would be more advanced than the current high-orbit infrared satellite reconnaissance and early warning by the US military.

Starlink can also enhance US precision strikes and command and control of unmanned equipment. With each Starlink satellite's bandwidth expected to reach 20 Gbps, this communication rate is sufficient to provide long-range communication services to US unmanned ground vehicles, robots, and drones deployed worldwide, strengthening the US military's long-range precision strike capabilities.

This will fundamentally transform the nature of future warfare.

Finally, Starlink can also transform into a space weapon, detecting and targeting space-based objects, akin to the "Brilliant Pebbles" space weapons born from the US "Star Wars" program. Starlink satellites are capable of emitting omnidirectional beams, allowing for telemetry, tracking, and control of spacecraft. This could turn into a high-precision system for calculating, simulating, and predicting rocket/missile trajectories, providing informational support for subsequent interception efforts. Starlink's automatic orbit alteration and avoidance capabilities, combined with 42,000 satellites in orbit, could also enable direct collision interceptions of intercontinental ballistic missile warheads.

Now, as per Musk's plans, the total number of Starlink satellites will reach 42,000, completely covering every corner of the Earth. The military capabilities hidden within the "Starlink Empire" warrant the attention of countries worldwide.

7.3   Musk's Ambition behind Starlink

As the CEO of SpaceX and the initiator of the Starlink project, Elon Musk's ambition is increasingly evident as he gradually launches 42,000 satellites into space.

Indeed, Starlink and aerospace are just one aspect of the commercial empire Musk aims to build. Concurrently with Starlink, Musk is also making continuous strides in other cutting-edge technological fields, including autonomous driving, AI, social media, brain-computer interfaces, and energy. What kind of game is Musk playing, and what can we discern from his entrepreneurial trajectory?

7.3.1   An Astonishing Entrepreneurial Journey

Musk seems to possess an innate propensity for adventure. He believes five areas will profoundly impact humanity's future: the Internet, new energy, space exploration, AI, and life sciences. While excelling in just one field is challenging enough, Musk has, over decades, incredibly ventured into these five areas and achieved remarkable success.

From 1995 to 2000, Musk and his partners founded three companies: Zip2, an online content publishing software; X.com, an electronic payment company; and PayPal, which evolved from X.com. Through acquisitions, management conflicts, and departures, Musk achieved financial independence and accumulated experience in company operations, deepening his understanding of the Internet.

The subsequent ventures of SpaceX, Tesla, and SolarCity are the main acts in Musk's career. In June 2002, Musk invested $100 million to found SpaceX, serving as CEO and CTO. From its inception, Musk stated SpaceX's goal to reduce rocket launch costs and plan for future Mars colonization, aiming to build a true space civilization.

At the time, these ideas were considered the dreams of a madman. In fact, as early as 2001, Musk had planned a project called "Mars Oasis," aiming to establish a small experimental greenhouse on Mars to try growing Earth crops in Martian soil.

In March 2004, Musk invested $6.5 million of his PayPal earnings out of a total of $7.5 million in Tesla's Series A funding. Musk undoubtedly became Tesla's chairman and largest shareholder. Initially, Tesla was founded on July 1, 2003, by two Americans as an electric vehicle and energy company. Musk's investment in Tesla came with the condition of becoming the company's chairman and having final decision-making authority. Three years later, Tesla's founder and first CEO, Eberhard, was ousted by Musk. Today, Tesla is indelibly marked by Musk's personal imprint, overshadowing its original founders. Musk truly brought Tesla into the world. In June 2010, Tesla went public on NASDAQ, earning Musk $630 million on its opening day. Tesla now comfortably leads the global electric vehicle market, accounting for about 30% of total sales.

Musk's achievements in rocket launches and electric vehicles are evident not only in the profitability or market performance of these companies but also in the stunning products they've produced. SpaceX's independently developed Falcon Heavy rocket achieved a "single rocket, 60 satellites" launch break-through, while Tesla cars genuinely solved the shortcomings of other pure electric vehicles and traditional gasoline cars in terms of battery life, accelera-tion, and handling.

Success, however, has not been without struggles. In 2006, SpaceX's first rocket, Falcon 1, failed on its maiden flight, spinning out of control and crashing into the sea 25 seconds after liftoff. In 2007, Falcon 1 failed to launch again, unable to reach the required speed. On August 3, 2008, Falcon 1's third launch, carrying three satellites and the ashes of hundreds of space enthusiasts, began trembling two minutes into flight before losing contact.

These three consecutive failures dealt a fatal blow to Musk and SpaceX. His personal fortune dwindled, barely enough for one last launch. Amid the financial crisis, Tesla also faced bankruptcy due to product issues, with employees leaving en masse. Musk was on the brink of a breakdown, recalling 2008 as the "darkest year of his life."

Finally, just over a month after the third failure, SpaceX launched Falcon 1 again, successfully sending the first non-government rocket into space. This success brought SpaceX a NASA contract, ensuring its survival.

From SpaceX and Tesla, Musk's business empire accelerated: In 2006, Musk invested $10 million in SolarCity, a photovoltaic power company, now the largest commercial solar panel installer for consumers and businesses. Despite controversy over Tesla's acquisition of SolarCity, Musk's vision of a "globally unique vertically integrated energy company offering end-to-end clean energy products" is undeniable. Musk also founded The Boring Company in December 2016 to address surface congestion.

In 2016, Musk founded Neuralink, focusing on brain-computer interface technology, enabling direct communication between humans and machines without a physical interface. In July 2019, Musk revealed Neuralink's progress, describing a "sewing machine-like" robot capable of implanting ultra-fine threads deep into the brain, eventually reading and writing vast amounts of information. By May 2023, Musk's brain-computer interface product had received FDA approval for human trials.

Musk's latest significant move was acquiring Twitter (now renamed "X") in late 2022. After months of public sparring, Musk completed the purchase, becoming the owner of one of the world's most famous social platforms for $44 billion. Acquiring Twitter also filled a crucial piece in Musk's business puzzle—Internet media.

From SpaceX to Starlink, from Tesla to the Hyperloop, from brain-computer interfaces to virtual public opinion arenas, Musk's trajectory reflects an "unthinkable" logic beyond ordinary business comprehension.

7.3.2 What Is Musk's Next Move?

It is evident that the industries of Musk's companies are all forward-looking, cutting-edge technological fields, with ventures like Starlink and Tesla achieving undeniable commercial success. So, what can we glean from Musk's entrepreneurial trajectory? What kind of game is Musk playing?

Musk's business empire arguably harbors greater ambitions than Apple, given his approach to integrating terrestrial and space technologies. Within Musk's industrial framework, electric vehicles, new energy, AI, brain-computer interfaces, and Internet media are all interconnected in the short or long term, gradually forming an industrial closed loop, with Starlink being the most critical technology permeating all industries.

Today, we know that the most competitive communication technology in the field is not 5G networks, but satellite-based communication, especially Starlink, which can achieve broader coverage and establish interstellar communications. Although Starlink's advantages are not yet apparent, they will become more pronounced as performance continues to improve, reception technology becomes more miniaturized, user base expands, and usage costs decrease.

First, Tesla is likely to be the first to achieve autonomous driving, primarily due to Starlink's satellite communication. Today's 5G communication still relies on ground stations, and any system dependent on stations will face communication latency during signal switching and uneven signal coverage, which could pose fatal risks for autonomous driving.

In contrast, Starlink can provide relatively equal coverage without the latency issues associated with station switching. Starlink's future could enable Tesla to achieve more reliable and efficient data transmission, driving the realization of true autonomous driving. This new communication strategy allows Tesla to maintain a stable communication link even under high-speed travel and uneven signal coverage, which is crucial for achieving autonomous driving, particularly when users demand higher safety and stability. Musk's entry from the communication link, connecting hardware and software, could build a robust ecological closed loop.

However, cars are just one part of Musk's hardware ecosystem in his empire, and arguably the most critical part at present, with the strongest application dependency. Subsequently, Musk is likely to expand related hardware products around his business empire ambitions, such as flying cars.

Musk has invested in the flying car company Alef Aeronautics, and in 2023, Alef Aeronautics' electric flying car Model A received legal flight permission in the US. Musk is poised to become a creator of the flying car era. With Starlink's global network coverage, the era of flying cars will move from science fiction to reality.

In Musk's vision, flying cars are not just a means of transport but a revolutionary technology to solve traffic issues. Flying cars can provide fast, efficient, and low-carbon travel, changing the congestion and emissions problems of traditional transportation. Their advent will bring revolutionary changes to urban transport, greatly improving efficiency and comfort.

The key to flying cars, besides autonomous driving technology and precise navigation systems, is synchronous latency. Here, Starlink communication, with its high speed, low latency, and broad coverage, provides an ideal communication infrastructure for flying cars. Starlink's global network ensures efficient communication for flying cars everywhere, enabling real-time information transfer and coordinated operation. This feasibility opens the door for the widespread application of flying cars, laying a solid foundation for Musk's vision. Supported by Starlink communication technology, flying cars are set to become an integral part of the future transportation system, offering a more convenient, efficient, and innovative travel experience.

Then there's the need for applications to complement the hardware, which is why Musk is venturing into Internet media. The real purpose behind Musk's acquisition of Twitter is not the platform itself but its user base. After acquiring Twitter, Musk will likely transform it, given that its current model is somewhat outdated. Musk's acquisition has had a significant impact on Facebook, directly challenging Meta at the social level.

Acquiring Twitter is just one move in Musk's business empire. In the future, Musk is likely to develop an ecosystem through acquisitions, partnerships, or his own developments, possibly along with Musk's smartphone, to build a closed-loop ecosystem similar to, but more powerful than, Apple's.

Once Starlink communication is completed, Musk can miniaturize Starlink's communication receiver and implant it into his smartphones, electric cars, and all his smart hardware products. This move not only challenges Apple's business empire but poses a significant challenge to many current enterprises. To join Musk's ecological closed loop, we will first need to access and obtain authorization for his Starlink technology.

7.3.3 The New Overlord of Artificial Intelligence

Apart from providing a global communication network, Starlink also contributes to computational breakthroughs, propelling Musk to become the new overlord of AI.

With the breakthroughs in AI, exemplified by ChatGPT, AI is entering an unprecedentedly rapid development phase. Concurrently, computational power has become an increasingly significant limitation for AI's advancement.

In the era of classical physics, computational power was mainly provided by leading chip processors. However, an equally important technology was processor cooling, which typically requires a considerable amount of electricity and cooling infrastructure. This presents a significant challenge in cloud service platforms, where the costs of electricity and cooling water usage, in addition to hardware purchases, form a significant economic burden and exert substantial environmental pressure.

Musk's Starlink, however, could revolutionarily address this issue, such as by situating servers in outer space. The core advantage of this approach lies in the naturally low-temperature environment of outer space, eliminating the need for additional cooling equipment, thereby significantly reducing the energy and cost burden of data center operations.

The Starlink satellite network could form a vast, distributed data center, with satellites acting as mobile servers, providing services to users on Earth anytime and anywhere. This distributed structure effectively eliminates the regional limitations and single-point failure risks faced by traditional data centers. Starlink's satellites serve both as mediums for data transmission and computational nodes, providing powerful computational support for AI super-computing.

Moreover, with advancements in quantum computing, Musk may further upgrade Starlink. After all, quantum computing, a disruptive technology, transcends the limitations of traditional computers and has the potential to process large-scale data instantaneously.

Therefore, through upgrading Starlink, Musk could build a satellite network not only capable of quantum computing but also equipped with quantum communication capabilities. This upgrade would transform Starlink from a traditional data transmission tool into a powerful computing platform. If successful, Starlink would become a satellite network with quantum computing capabilities, providing more efficient and advanced support for the computational demands of the AI era. This innovation would play a breakthrough role in scientific research and impact various sectors, including healthcare and finance. In healthcare, Starlink's quantum computing capability could accelerate the analysis of genomic data and the processing of medical images, offering more precise solutions for disease diagnosis and treatment. In finance, quantum computing could be used for more complex risk analysis and optimization of trading models, enhancing the efficiency and stability of financial systems.

In the long run, Starlink's potential is immense—it can cater to computational needs on Earth and provide advanced computational and communication support for space exploration and interstellar communication. Musk's business empire is set to expand further into the realm of AI.

7.3.4 When Will the Metaverse Be Realized?

In 2021, the concept of the Metaverse successfully broke into mainstream consciousness. However, two years later, it gradually faded from public memory. In the second half of 2022, the cryptocurrency market experienced significant turbulence, followed by a plunge in NFT trading volume, platform disappearances, and continuous devaluation of high-priced digital collectibles. The real sign of the Metaverse engine stalling was the retreating stance of pioneers in this field, such as Meta, named after the Metaverse.

The fading of the Metaverse was almost inevitable. In fact, what we discuss today as the Metaverse still remains largely at the level of science fiction creation. None of the underlying industrial chain technologies that genuinely support the construction of the Metaverse currently meet the requirements. The day the Metaverse becomes a reality, the era we currently call the Metaverse will likely be named differently, based on the new technologies and predictions of that future time.

When the Metaverse will be realized involves the integration of physical entity time with the digital virtual world, and interconnectivity and interaction facilitated by communication technology and intelligent sensing systems.

This highlights two key aspects of the realization of the Metaverse. First, achieving the intelligent transformation of all things and constructing a digital twin of Earth. Second, handling massive data volumes and achieving real-time information interaction, relying on advanced quantum communication and quantum computing technologies. Both these aspects need a common support system, precisely what a global satellite communication network like Starlink can provide.

Building a digital twin of Earth is foundational for realizing the Metaverse. A digital twin is a digital replica of the real world, achieved through digital modeling of various objects and processes on Earth, closely simulating the real world. This requires an abundance of sensors, cameras, and data collection devices, along with reliable communication and strong computational power to transmit, process, and analyze the collected data. Starlink, as a global satellite network, provides extensive and powerful support for constructing the digital twin Earth. Starlink's communication capabilities enable the interconnectedness of various parts of the digital twin, facilitating real-time data transmission and updates.

The massive data volumes and real-time information interaction in the Metaverse impose higher demands on communication and computation technologies. Traditional communication networks and computing systems, due to limitations in transmission speed and processing power, struggle to handle the complex and massive data flows in the Metaverse. The advent of quantum communication and quantum computing offers a solution.

Quantum communication uses the properties of quantum entanglement to achieve highly secure communication. Due to the unique nature of quantum entanglement, any interception of information immediately causes changes, making communication highly secure and reliable. In the Metaverse, this will help ensure the security of massive data flows during transmission, preventing information leakage and tampering. Quantum computing, based on the superposition states of quantum bits, can instantly process complex computational problems. In the Metaverse, this provides more efficient and rapid support for real-time data processing and information analysis. The Starlink satellite network, by providing quantum communication and connecting to quantum computing capabilities, will facilitate the rapid transmission and processing of massive data globally in the Metaverse.

In other words, if Musk can independently develop satellites with quantum communication and quantum computing capabilities and directly upgrade Starlink, it will open up broader application fields for Starlink, making it an essential pillar in building the Metaverse. If independent development proves challenging, Musk might consolidate Starlink's leading position in the Metaverse construction by acquiring satellite companies with quantum computing and quantum communication capabilities.

It can be said that the development of Starlink is impacting the realization of the Metaverse. Through the construction of a digital twin Earth and the application of quantum communication and quantum computing technologies, Starlink is poised to become the link connecting the real world with the virtual digital world, accelerating the rapid development of the Metaverse.

7.3.5　Building the Era of Web 3.0

Today, many are envisioning the advent of the Web 3.0 era. The core concept of Web 3.0 lies in decentralization, i.e., personalization of data sovereignty. Web 3.0 will truly propel humanity into an era of data sovereignty, safeguarding individual data rights and realizing personal data entitlements.

It's important to note that current tech giants like Google, Apple, and Amazon heavily rely on centralized data sovereignty in their business models, appropriating user behavioral data for processing, analysis, and application in various commercial activities. However, Musk, leveraging the advantages of Starlink communication, could establish a new model of personalized data sovereignty and disrupt the centralized business landscape.

Starlink communication is characterized by decentralization. Its unique satellite network architecture differs from traditional centralized communication systems, employing satellites distributed in Earth's orbit to form a vast and dispersed network. This design does not depend on a single central node but allows multiple nodes to participate in data transmission and communication tasks, laying a solid foundation for the realization of Web 3.0. Traditional data centers and communication networks, often controlled by a few companies or institutions, centralize data sovereignty in these organizations. In contrast, Starlink's decentralized network structure disperses data across satellites globally, making data more scattered, secure, and less prone to control and misuse by single entities.

Based on Starlink, Musk could establish a system of personal data exchanges. In this new system, individuals would own their data sovereignty, deciding whether to share their data and with whom. Starlink's efficient communication network and decentralized nature enable individuals to participate more directly in data trading and circulation, free from a few centralized platforms. This would help build a fairer and more transparent data trading ecosystem, enhancing users' control and rights over their data.

Moreover, Starlink's technical features could provide more secure and private data protection. Traditional centralized data platforms pose risks of data leakage and misuse. In contrast, Starlink's use of encryption technology and decentralized storage can enhance user data security and reduce the risk of unauthorized access, laying a solid foundation for a privacy-protected data sovereignty system.

Leveraging Starlink's advantages, Musk could promote the establishment of a new model of personalized data sovereignty in the Web 3.0 era. Decentralized communication networks, personal participation in data exchanges, and more secure data protection could offer individuals more data sovereignty choices, leading humanity into a fairer, more transparent, and individual rights-protected data era.

In the future, under Starlink's seamless coverage, every Tesla or flying car equipped with solar power will connect with and to the world. Entering the Metaverse era, humans capable of high-level physical and biological human-machine interaction will be intelligently connected to everything in the world, while owning data sovereignty, controlling, and trading their data. Above the satellites, a spacecraft exploring farther into space will head toward Mars. Ultimately, Musk's goal is to construct and shape a "super-intelligent state integrating heaven, Earth, humans, and objects"—a future existence centered on humans, connecting Earth and beyond, achieving intelligent connection, seamless interaction, and wisdom sharing between heaven, Earth, humans, and objects.

EPILOGUE

Undoubtedly, we are living in a grand era. Technological innovations are accelerating our entry into a science-fiction age at a rate unparalleled in any previous century. Technologies like satellite communication, AI, the metaverse, quantum computing, autonomous driving, space travel, and brain-computer interfaces, each have the potential to transform human society.

These technologies are not developing in isolation but are closely interconnected. The development of one technology catalyzes the maturity of another, leading to a foreseeable explosion of technological advancements. For example, with the rise of ChatGPT, AI has entered a new phase of rapid development and application, yet it is simultaneously limited by computational power. Quantum computing, capable of breaking this bottleneck, is crucial for AI's further leap.

Among these transformative technologies, one stands out as the "connector" and "enabler" of all others: satellite communication.

Currently, quantum computing, a revolutionary technology, is emerging, offering boundless potential for computational power. In traditional computing, information exists in the form of bits. Quantum computing, however, utilizes the superposition and entanglement properties of quantum bits, enabling computers to process multiple potential states simultaneously. This parallelism dramatically exceeds traditional computing in solving complex problems, enhancing speed and efficiency.

In this trend, chips are also moving toward quantum technology. The quantum transformation of chips, combined with AI, will usher in true endpoint intelligence and computational power, marking the real beginning of the era of edge computing. Endpoint intelligence means future sensors will transcend simple monitoring or digitization of monitored objects and then relay information to a cloud center for processing. Instead, they will possess powerful quantum computational capabilities and AI decision-making abilities.

Endpoints will become intelligent centers with self-computational abilities, rather than mere peripheral monitoring devices. This transformation will elevate sensor technology to new heights, shifting the core of information processing from cloud centers to endpoints for faster, smarter decision-making.

In this intelligent era, both horizontal and vertical integrated intelligent communication systems will be key. Horizontal communication refers to direct communication between intelligent sensors. In the future, leveraging the computational power of quantum chips, AI's intelligent decision-making, and 6G communication technology, endpoint devices will directly realize real-time communication, interaction, and decision-making. Vertical communication is the direct interaction of all endpoint devices with a satellite communication center, building a larger, more comprehensive intelligent command and control system. This centralized AI control system will facilitate real-time management and dispatch of various endpoints, providing a technological foundation for integrated aerial and terrestrial transportation systems like flying cars and autonomous vehicles.

The Starlink communication system will play a pivotal role in this increasingly intelligent ecosystem. Its decentralized nature makes horizontal communication smoother, enabling direct interconnectivity between intelligent endpoints. Simultaneously, as a communication center, Starlink will powerfully support vertical communication, seamlessly integrating various intelligent endpoints with a centralized AI system. This comprehensive three-dimensional intelligent communication system will advance intelligent transportation systems, smart city management, and other fields.

In fact, the essence of the metaverse, as I once proposed, is the digitization of everything. That is, everything on Earth, including humans, becomes interconnected and interactive in a new form after being equipped with intelligent sensors, creating a virtual digital twin Earth that interacts with the physical Earth. Through intelligent wearable sensors, humans can constantly sense and record various physiological indicators, emotional states, and behaviors. This data not only provides a comprehensive understanding of individual health and psychological status but also offers rich resources for scientific research in medicine, psychology, and other fields. Simultaneously, everything else on Earth will be digitized. The natural environment, flora and fauna, buildings, etc., will become digital twins in the metaverse, constantly reporting information through intelligent sensors, achieving all-around monitoring and management. This will provide unprecedented data support for environmental protection, resource management, and urban planning.

The construction of this metaverse form is crucially dependent on intelligent sensor-based technology and satellite communication technology developed by Starlink. Starlink's decentralized communication system will facilitate both horizontal and vertical communications. Intelligent sensors can directly communicate horizontally, forming a vast self-organizing network for real-time interconnectivity between devices. Vertical communication, through Starlink, connects each endpoint with a centralized AI system, enabling real-time management and dispatch of each intelligent endpoint.

This integrated horizontal and vertical intelligent communication system will greatly propel the development of the metaverse. Endpoint devices can communicate and interact directly, achieving more intelligent, autonomous decision-making. Meanwhile, the centralized AI system can monitor and coordinate all endpoints in real-time through Starlink, providing efficient management for the entire system. This will support the realization of integrated aerial and terrestrial transportation systems, intelligent city management, and other scenarios.

In the future, the construction of the metaverse will establish a new form of interconnectedness and interaction between the virtual digital twin Earth and the physical Earth. Through intelligent wearable sensors and Starlink communication technology, people will be able to seamlessly interact between the virtual and real worlds. This will not only change the way we perceive the real world but also create new virtual reality experiences, thereby promoting developments in entertainment, social interaction, education, and more. For instance, in education, the metaverse will enable truly unrestricted real-time online learning. With intelligent sensors and satellite communication, students can enter virtual reality classrooms from anywhere in the world for immersive learning. Whether on a cruise ship at sea, Mount Everest, the poles, or space, learners can participate in real-time learning through satellite communication. This approach will completely eliminate the inequality of educational resources between regions and countries. By accessing the educational space of the metaverse, everyone can enjoy high-quality learning resources.

In healthcare, wearable medical monitoring devices will allow individuals to monitor their health and nutritional indicators in real-time. With the aid of AI doctors, they can receive real-time monitoring results and health risk alerts, along with adjustment recommendations. In homes, smart wearable devices could detect sudden conditions like heart attacks or strokes at night, prompting home humanoid robots to perform emergency aid. Satellite communication can also establish real-time contact with hospitals, coordinate emergency resources, and improve medical treatment efficiency.

Beyond education and healthcare, the widespread application of satellite communication will modernize and intelligentize all aspects of human life and production. This leads humanity toward a smarter, more efficient, and interconnected future. With the rapid development of satellite communication technology, an unprecedented era of science fiction is approaching. A digitized, intelligent twin Earth is coming our way. In this epochal era of technological transformation, learning and innovation are the only keys to staying in step with the times.

REFERENCES

Central Securities Mobile Communication Technology Co., Ltd. *2023 White Paper on Satellite-Ground Fusion Communication.*

Dongfang Securities. *Industry Study of Satellite Internet: Key Elements of 6G Integrated Network in Sky, Space, and Earth.*

Fang, Min, Duan Xiangyang, and Hu Liujun. "Challenges, Innovations, and Prospects of 6G Technology." *ZTE Communications Technology* 26, no. 03 (2020): 61–70.

Guosen Securities. *Deep Report on Satellite Communication Industry: Abundant Potential in Satellite Applications, Significant Opportunities in Operational Aspects.*

———. *Special Topic on Satellite Communication Industry: Acceleration of Satellite Internet, Focus on Opportunities in Low-Earth Orbit Satellite Industry Chain.*

He, Kang. "Starlink: Reconstruction of Global Satellite Internet Era Communication System." *Journal of Hunan University of Technology (Social Science Edition)* 25, no. 04 (2020): 23–31.

Li, Rui, Lin Baojun, Liu Yingchun, et al. "Overview of Laser Inter-Satellite Link Development: Current Status, Trends, and Prospects." *Infrared and Laser Engineering* 52, no. 03 (2023): 133–147.

Li, Zhen, Bashir Ali Kashif, Yu Keping, Al Otaibi Yasser D., Heng Foh Chuan, and Pei Xiao. "Energy-Efficient Random Access for LEO Satellite-Assisted 6G Internet of Remote Things." *IEEE Internet of Things Journal* (2021).

Liu, Jia. "Analysis and Prospects of Low Earth Orbit Broadband Satellite Communication Industry." *China New Communications* 23, no. 01 (2021): 108–110.

Liu, Xuguang, Qian Zhisheng, Zhou Jihang, et al. "Reflections on the Satellite System of 'Starlink' and the Development of Domestic Satellite Internet Constellation." *Communication Technology* 55, no. 02 (2022): 197–204.

REFERENCES

Liu, Yu. "The Battle between 'Beidou' Second Generation and 'Galileo' Satellite Frequencies." *Digital Communication World*, no. 02 (2011): 16–17.

Liu, Zheming, Wu Yunfei, Wei Xiao, et al. "Application Scenarios, Architecture, and Key Technical Challenges of 6G Satellite-Ground Fusion Networks." *Wireless Communications Technology* 48, no. 06 (2022): 1058–1064.

McDowell, Jonathan C. "The Low Earth Orbit Satellite Population and Impacts of the SpaceX Starlink Constellation." *The Astrophysical Journal Letters* (2020).

Miao, Deshan, Chai Li, Sun Jiancheng, et al. "Research and Evolution Outlook of Key Technologies for 5G NTN." *Telecom Science* 38, no. 03 (2022): 10–21.

SpaceX. "SpaceX Starlink Factory Building Satellites Four Times Faster than Closest Competitor." 2020. Available at https://www.teslarati.com/spacex-starlink-satellite-production-faster-than-competitors.

Sun, Yaohua, and Peng Mugen. "Low Earth Orbit Satellite Communication for Direct Connection to Mobile Phones: Key Technologies, Current Developments, and Future Prospects." *Telecom Science* 39, no. 02 (2023): 25–36.

Wang, Taijun, Tang Shiqi, and Zhou Chao. "Exploration of the Application of 'Starlink' in the Military Conflict between Russia and Ukraine." *Communication Technology* 55, no. 08 (2022): 1006–1013.

Wu, Qilong, Long Kun, and Zhu Qichao. "International Competition in the Field of Low Earth Orbit Satellite Communication Networks: Situation, Motivation, and Participation Strategy." *World Science and Technology Research and Development* 42, no. 06 (2020): 587–597. doi:10.16507/j.issn.1006-6055.2020.12.004.

Wu, Xiaowen, Jiao Zhenfeng, Ling Xiang, et al. "Prospects of Satellite Communication Network Architecture for 6G." *Telecom Science* 37, no. 07 (2021): 1–14.

Yu, Nanping, and Yan Jiajie. "The US 'Starlink' Plan and Its Impact from the Perspective of International and National Security." *Journal of International Security* 39, no. 05 (2021): 67–91, 158–159. doi:10.14093/j.cnki.cn10-1132/d.2021.05.003.

Zhai, Hua. "Analysis of the Integration of 5G and Satellite Mobile Communication Systems."

Zhang, Jiayi. "Review of the Development of 'Starlink' System in 2023."

Zhang, Linfeng. *A Brief History of Communication: From Pigeons to 6G+*.

Zhang, Yulin, et al. *Satellite Constellation Theory and Design*.

Zhang, Zhengquan, Xiao Yue, Ma Zheng, Xiao Ming, Ding Zhiguo, Lei Xianfu, George K Karagiannidis, and Fan Pingzhi. "6G Wireless Networks: Vision, Requirements, Architecture, and Key Technologies." *IEEE Vehicular Technology Magazine* (2019).

Zhao, Qiuyan, Hu Zhaobin, Chen Chuan, et al. "Opportunities and Challenges of Large-Scale Low Earth Orbit Constellation." *Space Debris Research* 20, no. 01 (2020): 1–9.

Zhao, Xiaoling, Luo Yuwei, Wang Jianwei, et al. "Research on the Application Industry Chain of the US Starlink Project." *Satellite Application*, no. 10 (2022): 53–57.

Zheng, Wei. "The Balance of Profits and Losses behind Starlink: The Satellite Communication Race Is Urgent."

Zheshang Securities Co., Ltd. *Special Research on Satellite Communication Industry: China's "Starlink" Is Rising.*

Zhongxin Securities. *Industry Research Report on Satellite Internet: Accelerated Development of Satellite Communication, Promising Future of Integrated Space-Ground Communication.*

INDEX

ABOUT THE AUTHOR

Kevin Chen is a renowned science and technology writer and scholar. He was a visiting scholar at Columbia University, a postdoctoral scholar at the University of Cambridge, and an invited course professor at Peking University. He has served as a special commentator and columnist for the *People's Daily*, CCTV, China Business Network, SINA, NetEase, and many other media outlets. He has published monographs in numerous domains, including finance, science and technology, real estate, medical treatments, and industrial design. He currently lives in Hong Kong.